The Environmental Crisis

The Environmental Crisis
Understanding the Value of Nature

Mark Rowlands
Lecturer in Philosophy
University College
Cork
Ireland

Consultant Editor: Jo Campling

First published in Great Britain 2000 by
MACMILLAN PRESS LTD
Houndmills, Basingstoke, Hampshire RG21 6XS and London
Companies and representatives throughout the world

A catalogue record for this book is available from the British Library.

ISBN 0–333–74896–4

First published in the United States of America 2000 by
ST. MARTIN'S PRESS, INC.,
Scholarly and Reference Division,
175 Fifth Avenue, New York, N.Y. 10010

ISBN 0–312–23235–7

Library of Congress Cataloging-in-Publication Data
Rowlands, Mark.
The environmental crisis : understanding the value of nature / Mark Rowlands.
p.   cm.
Includes bibliographical references and index.
ISBN 0–312–23235–7 (cloth)
1. Environmental ethics. I. Title.

GE42 .R69   2000
179'.1 — dc21
99–059239

This book is printed on paper suitable for recycling and made from fully managed and sustained forest sources.

10   9   8   7   6   5   4   3   2   1
09   08   07   06   05   04   03   02   01   00

Printed and bound in Great Britain by
Antony Rowe Ltd, Chippenham, Wiltshire

Contents

Preface and Acknowledgements

There is, I think, a very real sense in which philosophy has failed the natural environment. Given the framework within which modern philosophy has unfolded, it is virtually impossible to make any sense of the claim that the environment has value in and of itself. To be sure, we can understand, with little difficulty, the idea that the environment is valuable because of what it can do for us – tectonically, economically, scientifically, medically, recreationally, even spiritually – because we are valuers of the environment, and we are comfortable with the idea that the value of an item can be a function of its being valued by a valuing consciousness. And many of us can understand how the environment can have value relative to valuers who are not human. Many non-human animals, it seems overwhelmingly likely, are valuers akin to ourselves, and so there is no great theoretical or conceptual difficulty in understanding how the value of the environment can be a function of their evaluative acts also. But where modern philosophy almost inevitably, and perhaps necessarily, draws a blank is in understanding how the environment can have value in itself, of itself, and independently of valuing consciousness.

In the eyes of many, and almost certainly most, this is a strength and not a weakness of modern philosophy. Philosophy draws a blank in understanding how the environment can have value in itself simply because there is nothing to understand here. The idea that there can be value independently of valuing consciousness is confused, and probably incoherent. And this, I think, is a claim with which it is impossible to disagree. Given the framework for understanding value bequeathed us by philosophy, the idea of *intrinsic* value, in this sense, is indeed a confused one. The sense in which philosophy has failed the environment, then, lies not in that it denies the existence of a spurious form of value. Rather, it lies in the framework which makes us see this spurious form of value as the only alternative to the claim that the value of the environment is anthropogenic.

What causes us to think of our options as constrained by the anthropogenic/intrinsic opposition is a deeper, more basic, opposition: that between *mind* and *world*. Accordingly, some environmental philosophers have urged the dismantling of the subject/object distinction; the breaking down of the distinction between mind-constituted and world-constituted reality. And with this I wholeheartedly concur. However, in environmental circles, what is not commonly recognised, indeed, as far as I can ascertain, universally unrecognised, is that there are two ways of dismantling this opposition, ways that are crucially different for the prospects of developing an account of the value of nature.

The most obvious gambit derives from Kant, although its roots, I think, lie much deeper. This strategy involves, as I put it, *pulling the world into the mind*: showing that what we regard as the world is, in fact, constituted by the structuring activities of the mind. This is the strategy of idealism, of neo-Kantianism broadly construed. Or, as I refer to it in this book, *humanism*. We are all familiar with humanism in this sense. Growing up in a neo-Kantian milieu, most of us cut our philosophical teeth on the intricacies and problematicity of the mind/world distinction, and its primary/secondary, fact/value, theory/observation offshoots. However, what goes almost universally unnoticed in this neo-Kantian age, is the existence of a quite different way of dismantling the mind–world distinction, a strategy that proceeds by, as I put it, *pulling the mind into the world*.

The fundamentals of this strategy emerged from work I had been doing in the philosophy of mind, work that has now spanned several years. In my earlier book *The Body in Mind: Understanding Cognitive Processes* (Cambridge University Press, 1999), I developed what I there referred to as an *environmentalist* model of cognition. In developing this model, I came to understand that it had ramifications not just for the understanding of cognition, and of the mind in general, but that it also had something to say about the nature of philosophy. The environmentalist model of cognition, if taken seriously, would subvert, undermine rather than refute, the humanist turn in philosophy, understood as the idea that the existence and nature of the world are dependent, to a greater or lesser extent, on the activities of the mind.

I came to believe that if we began from this sort of neo-Kantian humanist position, we would quickly find ourselves constrained, and constrained quite severely, in the conceptions of *value* we found available to us. If the existence and nature of the world are derivative upon the structuring activities of the mind, then it is difficult to avoid the conclusion that the value of the world is also thus derivative. If the world has secondary ontic and epistemic status, it is difficult to see how it does not also have secondary axiological status. The essentially Kantian milieu in which we have all grown up, that is, precludes the development of anything other than a secondary and derivative conception of the value of the environment. In this sense, the milieu fails the environment.

The key to understanding the value of the environment is not simply to break down the distinction between mind and world, between subject and object, a claim endorsed by several writers, but also to do it in the right way. And to do it in a way that makes the world ontically, epistemically, and hence axiologically dependent on the mind is emphatically not the right way. The purpose of this book, then, is threefold: (i) to show why, for the purposes of developing a satisfactory environmental ethic, we need to break down the subject/object distinction, (ii) to show how to do this in the right way and (iii) to develop a resulting conception of the value of the environment based on this dissolution. Chapters 2–4 are addressed to (i). They argue that traditional conceptions of environmental value, conceptions predicated on the mind/world distinction are inadequate. Chapters 5–8 are concerned with (ii), they show how to break down the subject/object distinction in a way that is not inimical to environmental thought. Chapter 9, directed at (iii), develops an account of the value of the environment on this basis. Chapter 10 feeds this back in to provide a perspective on the titular subject of this book: the environmental crisis.

The philosophical debts I have incurred over the years are too numerous to recount. But two, perhaps, are most relevant here. It was Colin McGinn, the only other philosopher-surfer I am aware of, who first got me obsessed about the location of the mind, an obsession that is becoming perhaps disturbingly prolonged, and whose most recent expression is to be found here. Thanks Colin.

In environmental philosophy, my greatest debt is undoubtedly to J. Baird Callicott. Someone who reads this book superficially might suppose that Callicott figures in it as an enemy. I hope it is clear, however, to anyone who gives the book more than a cursory glance, that I single out Callicott for criticism as a mark of respect. Indeed, it is possible for me to attack Callicott on various points only because, I think, we agree on so much more. For fairly obvious reasons, I choose to accentuate the differences between our positions, but these could quite easily be relegated to the margins of the debate, leaving massive agreement in the centre (although Callicott may see things differently). More concretely, Callicott was kind enough to read an entire draft of this book, and his comments significantly improved the eventual result. My thanks to him.

My thanks, as always, to Jo Campling. And also to Annabelle Buckley and Karen Brazier at Macmillan and to Jo North for some excellent copyediting. Thanks to the students in my environmental philosophy class at University College, Cork. Also, thanks to the University of Iceland, where I was a visiting fellow for a period of 1998 and had a totally brilliant time.

Finally, philosophy is a fickle yet demanding discipline. Unlike perhaps any other discipline, we philosophers are at the mercy of ideas, and we have to follow wherever they lead. What I mean is, in philosophy you cannot just decide that one day you are going to do X. Because, in a very real sense, we do not do philosophy, philosophy does us. Suppose, for example, one had been given a research grant by an insightful and benevolent body to pursue a project on consciousness. Then if, by a somewhat tortuous route, with the details of which I shall not bore you, the research mutates (by way of a study of Sartre's model of consciousness and his idea of a 'radical reversal of idealism' actually) into research into the value of the environment, then one should not, I think, be penalised with respect to future applications for research fundage. Philosophy is like that. A philosopher is like a lone sailor, out on the ocean. He/she is at the mercy of the winds, and sometimes the winds come and sometimes they do not. But when they do, he/she has to follow them wherever they blow. And a book is a book, right? So, hugs and kisses to the Faculty of Arts Research Grants Committee at University College, Cork.

1

'Where the Danger Is Grows also That which Saves'

1.1 Two pictures

We might begin with two pictures, both of which are prominent talismans of a certain type of environmental consciousness. Consider, first, the Portuguese man-o'-war, *phrysalia*, whose 20-metre poisonous tentacles trail behind it, and whose sail floats, powered by the wind, drive it forward. *Phrysalia* is, it is now generally accepted, not an organism but a community. It is a community because each *zooid*, each of the individual components that makes it up, is derived from a complete multicellular organism. In effect, it consists of thousands of tiny individual animals, stitched together, sharing common purpose and common fate. *Phrysalia* should thus be assimilated to a colony of bees or ants, rather than to an individual organism. Each zooid is, of course, incapable of independent existence. And the nature and behaviour of each zooid cannot be understood in isolation from the whole of which it is a part. But the corresponding claims are both true for ants and bees also. And like bees in a hive, or ants in a colony, each of these animals knows its obligations and its place. Just as colonies of ants and bees have workers, *phrysalia* has gastrozooids whose function is to collect food. As ants and bees have soldiers, *phrysalia* has dactylozooids whose function is to protect and defend. And as ants and bees have a queen, *phrysalia* has gonozooids whose function is to reproduce.

Phrysalia renders the distinction between organism and community somewhat problematic. And the problem can be pushed further. For there is a clear sense in which things that are uncontroversially organisms are also communities. Every organism is a collective. An organism consists of millions of single cells, each with its own identity and purpose, but also dependent on the whole, incapable of existing

1

independently of the collective of which it is a part. A whale is just as much a collective as *phrysalia*; it's just that a whale is a collective of a million billion collaborating cells, whereas *phrysalia* is a collective of ensembles of cells.[1] But the distinction between cells and ensembles of cells is not one that carries much weight. For cells themselves are collectives: collectives forged by the symbiotic collaboration of different types of bacteria. Every cell in the human body, for example, contains a mitochondrium. But a mitochondrium is a descendant of a tiny bacterium that, through the vicissitudes of natural selection, became extremely accomplished at producing energy, and eventually came to contract out this ability, and therefore surrender its independence, in exchange for a less precarious life within the cell walls of our ancestors. And mitochondria are, in turn, collectives: collectives of chromosomes which are themselves collectives of genes. In human cells, chromosomes go about in teams of twenty-three pairs. But this in turn is a collaborative, hence collective, enterprise. They could, for example, go about individually; indeed, they do so in bacteria. In biology, the concept of a community does not just apply to ensembles of organisms. On the contrary, communities have unexpected depth. Organisms are themselves communities, and these communities are made up of still further communities. Each organism is a veritable Russian doll of community collectives. *Phrysalia* is, therefore, a community in more ways, and on more levels, than one might think.

This picture of organisms as communities provides a powerful organising vision for some – we might say *deeper* – strands of environmental thought. If individuals are themselves communities, indeed communities composed of communities, then the distinction between the individual and the collective effectively breaks down. Whereas the environment is naively thought of as a collective or community of individuals, with the demise of the distinction between organism and community, there is no conceptual obstacle to regarding the environment as a kind of individual, a type of *superorganism*. This essential idea is enshrined in the *Gaia hypothesis*: the hypothesis that the Earth either is, or at the very least functions as, a self-regulating organism.[2] If our thinking is guided by this picture, then we might well come to regard the individuality of human persons, their metaphysical distinctness and isolation, their, to borrow an expression from Leibniz, *monadicity*, as an illusion engendered by post-Enlightenment thought, an illusion that must be abandoned. We are no more separable from our environment, neither physically, functionally, nor indeed conceptually, than an individual zooid is from *phrysalia*, or a mitochondrium is from a cell. It is as if the Enlightenment has left us as individual zooids, somehow become detached from our home, and trying to understand ourselves without knowledge of from where it is we come and to what it is we belong. Our project of self-understanding is, thus, doomed to failure. Nonetheless our abortive attempt at self-understanding here might breed intellectual monsters. In particular, we might understand ourselves as individuals whose freedom of action is morally guaranteed by *rights*, and whose freedom can only morally, as opposed to physically, be limited by the rights of other individuals. But, again, this is an illusion born of the original deception. We have no rights independently of our place of belonging any more than the arm has rights independently of the body to which it belongs.

Thinking about the environment that is oriented by the picture of *phrysalia*, then, will lead to emphasis on the essential connectedness of human beings with the environment which is their home. Human beings are not essentially isolated egos, metaphysical monads that only contingently or accidentally happen to find themselves in a world. Rather, they are essentially, as Heidegger put it, *beings-in-the-world*. Our connection to our environmental home is absolute, and constitutive of our very identity as the particular beings we are, and constitutive of the moral entitlements we might possess. This way of thinking about the environment is, thus, holistically oriented. It is characteristic of a mode of environmental thought often called *deep ecology*.[3]

There is another picture, quite opposite in both intent and content, that has, I think, played an equally influential role in organising thinking about the environment. While the picture of *phrysalia* articulates a vision of human beings as essentially interconnected with their environment, and emphasises the holistic character of human–environment interactions, this second picture articulates a quite distinct vision; one which emphasises the apartness, separation, alienation of humans from their environmental home. This second picture is one of *cancer*. Cancer is the failure of cells to stop replicating. Cells that replicate in this way thrive at the expense of ordinary cells, since the replication of these is kept in check by various mechanisms. Thus, cancerous tumours, provided that they stay sufficiently generalised in appearance to metastasise (i.e. spread throughout the body), are, in the absence of drastic external measures such as chemical or radiation based attack, bound to eventually take over the body. Human beings, it is often claimed, are like a cancer, a blight. Just as cancer is caused by the rapid and uncontrolled explosion of cells in an organism, so there has been a rapid and uncontrolled explosion of human beings on Earth. Just as cancer transforms more and more bodily cells into its own image, so humans transform more and more of the world into their own image. The effects of the exponential growth of the human population in this century are, thus, analogous to the effects of cancer on a human host.

This picture of the human species as a cancerous tumour orients our thinking towards an aggressive and individualistic conception of human beings. Humans are essentially manipulators and exploiters of the environment, feeding off the world in the endogenously parasitic way characteristic of a cancer. The picture articulates a vision of humans as individualistic, isolated egos, whose primary mode of interaction with the world is one or another form of exploitation. The world is a pointillist horrorscape of monadic centres of imperialist aggression feeding on the body of a defenceless host.

So what are we? How should we see ourselves? Are we beings-in-the-world, or exploiters-of-the-world? Plain citizens of a community, or its not so benevolent despot? Is the world our home or our carcass?

At this point, we should take a step back. We should, I think, be reluctant to allow our thinking to be regimented by either of these two pictures. The way things are, more often than not, is just too slippery for pictures. I think Wittgenstein was probably correct, to a significant extent, in his view that many philosophical problems arise through slavish adherence to an organising picture or vision that makes us think that things have to be a certain way. The task of the philosopher, on this view, is to undermine the grip the picture has on us and, thus, show us that things do not, in fact, have to be that way at all. Whether or not this is true as a general picture of philosophy, or philosophical problems, I think it certainly is true in the case of much of our thinking about the value of the environment. And, in particular, the simple idea that, with regard to human nature, relating to the world in a holistic, communal, organismic way is good, while relating to it in an aggressive, exploitative, monadic way is bad, is just that: simple. Far too simple. The pictures, in their apparent polar opposition, blind us to the possibility that the two visions of human beings are far more intimately related than might be apparent. Perhaps the two pictures are in some way mutually dependent, in the sense that humans can be in-the-world, fully part of the ecological community as envisaged by the collectivist picture only because we are also cancerous exploiters of the world. Perhaps it is only because we are monadic manipulators of the world that we can be, in relation to that world, like an individual zooid in *phrysalia*. This is the idea that will be explored in this book.

1.2 The demon Descartes

In accordance with the general antipathy towards the individualistic, egoistic, and monadic conception of the self that underwrites the cancerous vision of human nature, the seventeenth-century French philosopher René Descartes looms large in contemporary environmentalist demonology.

There is a view of the mind that began its life as a controversial philosophical thesis, and then evolved into common sense. The view is both widespread and tenacious, not only as an explicit doctrine, but, even more importantly, in the clandestine influence it has on explicit doctrines of the mind. The philosophical thesis from which the view was born was spelled out by Descartes, and its association with him is sufficiently robust for it to be called the *Cartesian conception*.

According to the Cartesian conception, minds are to be assimilated to the category of substance. That is, minds are objects which possess properties. Indeed, minds can be conceived of as relevantly similar to other bodily organs. Just as the heart circulates blood, the liver regulates metabolism, and the kidneys process waste products, the mind *thinks*. According to official Cartesian doctrine, the major difference between the mind and these other organs is that the mind is a *non-physical* substance. The mind and brain are distinct entities, and while the mind may receive input from the brain, and in turn send information back to the brain, the two are nonetheless distinct. The brain is a physical organ operating on exclusively mechanical principles, the mind is a non-physical organ governed by principles of reason. And there is, Descartes thought, no prospect of deriving the latter from the former. This view of human beings as composed of two essentially different kinds of thing is known as *dualism*.

The Cartesian conception has been famously ridiculed by the philosopher Gilbert Ryle, as the myth of the *ghost in the machine*.[4] And it has been Descartes' decision to make the mind ghostly, i.e. non-physical, that has drawn the principal fire from dissenters. The dissenters' case, here, has largely been successful, and not many philosophers today would classify themselves as Cartesian in this sense. Ryle's expression, however, has another side. Not only is Descartes' mind a ghost, but it is one that is *in* a machine. This, in fact, was the principal source of Ryle's ire. But whereas the revolt against ghostly views of the mind has been overwhelmingly successful, criticism of the second aspect of Descartes' view has, until recently at least, been comparatively muted. Most theorising about the mind is now based on the assumption that the mind is physical. However, it is also true that such theorising has been, and largely still is, based on the assumption that the mind is an internal entity, that is located inside the skin of any individual that possesses it. The revolt against Cartesianism, that is, has been restricted to the first aspect of Descartes' view, to his dualism. The other aspect, Descartes' *internalism* has, until recently, been largely ignored.

Descartes' dualism and his internalism have, arguably, the same root: the rise of mechanism associated with the scientific revolution. This revolution reintroduced the classical concept of the atom in somewhat new attire as an essentially mathematical entity whose primary qualities could be precisely quantified as modes or aspects of Euclidean space. Macroscopic bodies were composed of atoms, and the generation and corruption of the former was explained in terms of the combination and recombination of the latter. Atomism is, then, mechanistic in the sense that it reduces all causal transactions to the translation, from point to point, of elementary particles, and regards the behaviour of any macroscopic body as explicable in terms of the motions of the atoms that comprise it.

It is widely accepted that Descartes' dualism stems, at least in part, from his acceptance of mechanism. The physical world, for Descartes, is governed by purely mechanical principles. He was, however, unable to imagine how such principles could be extended to the thinking activities that make up the human mind. Minds, for Descartes, are essentially thinking things and, as such, governed by principles of reason. But such principles, Descartes thought, are distinct from and not reducible to principles of mechanical combination and association. Rationality, for Descartes, cannot be mechanised. Each mind is thus a small corner of a foreign field that is forever non-mechanical, hence forever non-physical. Descartes' dualism, in this way, stemmed quite directly from his mechanistic atomism.

Of equal significance, however, is the connection between mechanism and internalism. Mechanistic atomism is, we might say, methodologically individualist. A composite body is reducible to its constituent atoms. And the behaviour of a composite body is reducible to the behaviour of these constituent atoms. Thus, if we want to explain the behaviour of a macroscopic body, we need focus only on the local occurrences undergone by its parts. This methodological individualism would also have some purchase on the explanation of the behaviour of human beings since we are also, in part, physical. It is, therefore, no surprise that minds became analogously and derivatively conceived of by Descartes, and his dualist descendants, in atomistic terms. A mind, for Descartes, is essentially a *psychic monad*.[5] Each mind is a discrete substance insulated within an alien material cladding. Just like any other atom, the mind could interact with the physical atoms of the body. But, crucially, and again just like any other atom, the essential nature of the mind was not informed or altered by this interaction. The rational nature of the mind is taken as an independent given, and its interaction with other atoms is extrinsic to this nature. The ghosts of this conception of the mind are very much with us today.

While Descartes' dualism is almost entirely discredited, his internalism is very much alive, and it is this internalism, this view of the mind as located inside the skin of thinking subjects, which creates the view of human beings as essentially individualistic and monadic entities, point lights of subjectivity in a world bereft of feeling or purpose. And this, it is thought, underwrites the view that the primary mode of transaction between a person and the world should be one of manipulation or exploitation. If I am *in here*, and everything *out there* is not me, then, barring any sentimental attachment on my part, why should I be in any way concerned to foster or care for it? Why should I not use it for my purposes, bend it to my will? Monadicity is, therefore, the handmaiden of machination. Or so this line of argument goes.

The Cartesian conception of the mind as essentially an internal organ of the body brings with it the distinction between *subject* and *object*. If the mind is something whose location is restricted to the confines of the thinker's body, then there is a firm distinction to be drawn between the mind as the subject of thought and the external or environmental objects that the mind thinks about. The relation of the mind to the world then becomes derivatively conceived of on analogy with the relation between body and world; in terms, that is, of the relation between *inside* and *outside*. Just as a process occurring in, say, my heart is a process occurring inside my body, a process that can be caused by certain things happening outside my body (a sudden scare causes an increase in heart rate, etc.), so too when I think about something, this thinking is something going on *inside* and it is about something existing *outside*. This outside can now include not just things existing in the environment, but also things that are on the inside of my body; I can think about my increase in heart rate, for example, and this is something that exists outside my mind. Nonetheless, the relation of thinking to what is thought about has been modelled on, extrapolated and derived from, the relation between body and world. And the foundations of this derivation lie in Descartes' internalism; in his viewing the mind as essentially an *inside*.

Viewing the mind as an inside and the environment as an outside makes it very difficult, if not impossible, to develop a satisfactory account of the value of the environment. The metaphor of inside and outside presents us with a stark choice. If we want to claim that the environment has value, then we are committed to claiming that this value derives either from the inside, from the activities of the human mind, or from the outside, that it is objectively present in the environment independently of those activities. If we opt for the former, then it seems that eventually we will be driven to the view that the environment does not *really* possess value; that whatever value it has derives only from the internal valuing activities of the human mind. Should these activities cease, the value of the environment would dissipate. Real value lies on the inside; it is possessed by the outside only to the extent that the inside is able to project it outwards. The value of the environment, on the former option, is, therefore, secondary and derivative, and this simply seems to be a disguised way of denying that the environment has genuine or intrinsic value. The latter option faces a different problem: it makes it completely mysterious how the environment could have value. If we want to claim that the environment, the *outside*, has value independently of the valuing activities occurring on the *inside*, then it seems we are forced into the extremely difficult job of explaining how, exactly, there can be value in the absence of valuing. How can something have value in the absence, or potential absence, of its being valued? What sort of thing is this value that it can exist in the absence of valuing? As we shall see, attempts to furnish these questions with convincing answers have, historically, been very unsuccessful.

Accounts that seek to ground the value of the environment in the valuing activities of the human mind are known as *subjectivist* accounts. These will be the subject of Chapter 4. Accounts that seek to make the value of the environment independent of such internal valuing activities are known as *objectivist* accounts. These will be the subject of Chapter 3. Both types of account, I shall argue, face serious problems. However, the coherence of both types of account ultimately rests on Descartes' internalism, and the subsequent conceptualisation of the relation between mind and world on the model of inside and outside. So, if we were able to reject this model, then this would open the possibility of an alternative way of understanding the value of nature, a way that is neither subjectivist nor objectivist. Happily, as I shall try to show in the second half of this book, Descartes' account is not only eminently questionable; it is, in fact, seriously flawed. This leaves the possibility of developing an alternative account of environmental value, an account that is neither subjectivist nor objectivist.

Where this book parts company with received environmental wisdom, however, is in the role it assigns to manipulation and exploitation. Manipulation of the environment, it is commonly thought, goes hand in hand with a monadic, internalist view of human nature. If only we realised that, with respect to the environment, we are like individual zooids in *phrysalia*, then the motivation for dealing with this environment in predominantly manipulative or exploitative terms would be undercut. Our essential interrelatedness with the biosphere would be revealed, and we would adjust both our view of ourselves and our dealings with the environment accordingly. However, I shall argue that while we are indeed like zooids in *phrysalia*, we are this way precisely because, and only because, we have evolved, just like every other creature, to be a manipulator and exploiter of the environment. Descartes' model is indeed misguided. We are beings-in-the-world, not isolated Cartesian egos. But what makes us beings-in-the-world is that we have evolved to be its manipulators and exploiters. The Cartesian model of the mind must go, and with it the dichotomy between subjects and objects, between inside and outside, but the way to rid ourselves of these conceptual hindrances is not by emphasising the connection between monadicity and manipulation, but by breaking this connection. It is precisely because we are like cancer in the liver, that we can also be like zooids in *phrysalia*.

1.3 Pulling the world in versus pulling the mind out: humanism and environmentalism

In recent environmental thought, a certain degree of consensus is beginning to converge on the claim that developing an adequate account of environmental value requires rejection of the so-called *modernist* world-view.[6] This is the world-view based on Descartes' conceptualisation of the relation between mind and world as a relation of inside to outside; of a subject that is essentially an *internality*, and a world that is essentially an object for, and hence external to, that internality. What is required for the purposes of developing an adequate theory of environmental value is the principled rejection of Cartesian internalism, and the subject/object, inside/outside, distinction based on it. And this is a claim with which this book wholeheartedly agrees.

What is not commonly recognised, however, is that there are two quite different ways of dismantling Cartesian internalism; two quite distinct methods of breaking down the subject/object distinction. In describing these, I shall continue to use the philosophically dangerous metaphor of inside and outside. I assume, with Wittgenstein, that such imagery is dangerous to the extent that it tempts us into philosophical confusion, and that one task of the philosopher is to resist such temptation. Nonetheless, one of Wittgenstein's central methodological insights was that such imagery is, in itself, harmless as long as we realise that we do not have to think about matters in the way in which the imagery tempts us to think about them. The task, then, is the essentially Wittgensteinian one of showing that we do not have to think about the mind, and consequently about value, in the way in which the imagery of inside and outside tempts us to think. And this task requires at least initial use of the offending metaphor.

The first, and historically prominent, way of breaking down the subject/object distinction is, in effect, to *pull the world into the mind*. This is the strategy of *idealism*. I shall use the term 'idealism' here in a deliberately broad way. Consequently, idealism manifests itself in a variety of forms. What is essential to idealism, what unites its various forms, is the idea that the world is, in one way or another, dependent on the mind for its existence or nature. And since this mind is almost always thought of as a human mind, idealism is a form of *humanism*. Heidegger has argued that the history of Western philosophy can be regarded as the history of the increasing idealisation, the increasing humanisation, of the natural world.[7] Essentially the same point is made by Nietzsche in *Twilight of the Idols*, in a passage aptly entitled 'How the "Real World" at last became a myth'.[8] According to Heidegger and Nietzsche, the constitutive connection between the mind and reality was firmly established by Plato. Plato identifies reality with that which is intelligible; to that which can be understood, in a suitably trained and cognitively equipped subject, by reason alone. Reality and intelligibility are, thus, intrinsically related. The connection is strengthened by Descartes, who makes the criterion of reality of a state of affairs its representation to a knowing subject with certainty. This trend to increasing idealisation of the world continues with Kant, undoubtedly its most influential exponent. The world in which we live, and of which we are aware, is a world essentially constructed by the mind, an edifice built up through the architectural proclivities and activities of the mind's forms and categories.

As I shall use the term, *humanism* is equivalent to *neo-Kantian idealism*. That is, 'humanism' denotes any view which sees the world in which we live, and of which we are aware, as constructed, at least in part, by activities of the human mind. This construction can be either immediate or mediate in character. It will be mediated when the mind structures the world through construction of something else (a language, a theory, etc.), where it is this something else that immediately structures the world. This is, of course, a very broad construal of neo-Kantianism. Indeed, construed in this way, it is difficult to think of anyone who is not a neo-Kantian, or any branch of philosophy that is not neo-Kantian. On this characterisation, in fact, neo-Kantianism precedes Kant by a considerable length of time. Locke, Berkeley and Hume were all, to differing extents and in differing ways, neo-Kantians. Moreover, neo-Kantianism encompasses even avowedly anti-Kantian thinkers. Nietzsche ('This world is the will to power and nothing else. And you yourselves are this will to power and nothing else besides' etc.) is a neo-Kantian on this construal, if we assume that willing is an activity of the human mind.[9] The linguistic turn in philosophy is plausibly viewed as a form of neo-Kantianism. Indeed, the linguistic turn comes from adding to the Kantian idea that the world is constructed by activities of the mind, one of two further claims: either that language determines the structure of thought, or that language mirrors the structure of thought. But, in either case, the basic Kantian emphasis on the structuring activity of the mind is preserved.[10] Similarly, mainstream twentieth-century philosophy of science is about as neo-Kantian as you can get, although, like most recent versions of neo-Kantianism, it emphasises the structuring activities of *theory*, as opposed to the mind. For recent neo-Kantians, theory seems to play precisely the same role in constituting the world as the activities of the mind played for Kant. But since theories are themselves mind-dependent entities, this is a variation upon, rather than a departure from, neo-Kantianism. And, of course, most recent continental philosophy is neo-Kantian to the core.

We live in a neo-Kantian age. Humanism, in this sense, is the defining intellectual movement of our time. The above potted history of philosophy is no doubt wildly procrustean. But, it is nonetheless true that, raised as we are in this sort of humanist, neo-Kantian, philosophical milieu, we are all familiar with the idea that subjects cannot be separated from objects. If the mind, immediately or mediately, enters into the very constitution of the world, then the world is not simply an object for the mind. Hence, subjects cannot be separated from objects, and facts cannot be separated from values. We are all familiar with the problematicity of the subject/object distinction. Most of us cut our philosophical teeth on these sorts of problems and issues. So, in one sense at least, the breaking down of the subject/object distinction is not an idea whose time has come, it is an idea that has been with us for a long time.

If this is so, why is environmental philosophy still wrestling with this sort of problem, still advocating a dismantling of Cartesian internalism that took place long, long ago? I want to suggest that the reason this is so is because while neo-Kantian humanism undermines the subject/object distinction, and the fact/value distinction predicated upon it, it does not do this in the right way, or for the right reasons. It is perhaps clear how profoundly anti-environmental this humanist rejection of the subject/object distinction is. According to neo-Kantian idealism, the environment is essentially, in one way or another, a construction of the human mind. This gives the mind an absolute ontological and epistemological priority. The nature of the world depends on, and derives from, the nature of the mind. And, therefore, the best way to study the world, at least from a philosophical perspective, is via study of the structuring activities of the mind.

Eugene Hargrove has detailed the many ways in which philosophy has failed to come to terms with the existence of the environment and, consequently, failed to develop any satisfactory account of environmental value.[11] Neo-Kantian idealism is surely philosophy's greatest failure in this regard. If one has grown up in a philosophical milieu that accords the world only a secondary ontic and epistemic status, it is difficult to see how that milieu can accord the world anything but a secondary and derivative axiological status. One can have, therefore, a considerable amount of sympathy with Holmes Rolston III's complaint that in recent environmentalist attempts to expunge the subject/object distinction, and to develop an account of environmental value on this basis, the subjectivists have 'won all the chips'.[12] Now, strictly speaking, Rolston's claim is inaccurate. If neo-Kantian humanism breaks down the subject/object distinction, it is difficult to see how this could provide a vindication of subjectivism. As Callicott points out, it would be more accurate to say that the subjectivists have been dealt out of the game.[13] Nonetheless, what underlies Rolston's complaint is, I think, an intuition that is quite sound. Neo-Kantian idealism might break down the subject/object distinction, but it does so, as one might say, in a fundamentally subjectivist way. Hence, it does so in a way that is fundamentally unsuited for the purposes of environmental ethics; in a way that, necessarily, can accord only secondary and derivative worth to the environment.

If we start from the perspective of humanism or neo-Kantian idealism, and this is where we do start, since this is the milieu in which we have, philosophically speaking, grown up, then small wonder we are having problems developing an adequate account of environmental value. We must reject the subject/object distinction, yes, and the fact/value distinction predicated upon it. But we must reject these distinctions in the right way, and for the right reasons. It is not enough to be post-modernists about the subject/object distinction. We must also be *post-humanists*.

Humanism, neo-Kantian idealism, breaks down the subject/object distinction by, so to speak, pulling the world into the mind; by construing the world as, in one or another way, a construction of the mind. The other way of attempting to break down the subject/object distinction, the method that will be explored in this book, is by, so to speak, *pulling the mind into the world*. This approach is profoundly anti-idealistic. The Cartesian conception sees the mind as essentially an *inside*, defined in relation to the objects of which it thinks, objects which are *outside* it. But suppose the mind were not a self-contained entity in this sense. Suppose these objects of which the mind thinks were literally constituents of the mind. When one, for example, thinks of an object, that object is not something external to one's thought, it is literally a *constituent* of that thought. If this were so, then the mind would not be an inside defined in relation to an outside. It would already be outside. But this is not in the neo-Kantian humanist sense that the outside is, in part, constructed by the mind. Rather, the mind is, in part, constructed by the outside. If this were so, then mental states and processes would be located in the environment as much as in the head. Thinking would be something we do not merely in the head, but, more fundamentally, something we do in the world. To suppose this, then, is to envisage a way of breaking down the subject/object distinction in a way that is precisely converse to that of neo-Kantian idealism. It is to pull the mind into the world.

If neo-Kantian humanism makes the environment mental, then our view of the mind envisaged above makes the mental environmental. In developing this idea, we would, instead of developing an idealist model of the world, be developing an environmentalist model of the mind. This is one of the tasks of the second half of this book. There, I shall argue that cognitive processes such as perceiving, remembering, reasoning, and thinking involve, at least in part, the manipulation of certain sorts of environmental structures, structures that bear information relevant to the cognitive task in question. Therefore, such processes have these environmental structures quite literally as constituents. These processes are central to the account developed because they are precisely the types of process that, according to humanist orthodoxy, are responsible for the construction of the world by the mind.

The project of the latter half of this book, then, is to develop what I shall call an *environmentalist model* of the mind. This is a model of mental processes that regards them as, at least in part, made up of environmental structures, and of the manipulation of these structures in appropriate ways. It is, thus, a model of the mental that identifies the environment as the primary location of mental processes, and which makes those processes occurring in the head just a special case, and not an especially interesting case, of this broader environmental location. This environmentalist model of the mental will then be used to develop a conception of environmental value that is both post-modern and, more importantly, post-humanist.

An environmentalist model of cognition is not in any way a model peculiar to the concerns of this book. It is certainly not a model dreamed up in an *ad hoc* way to suit the purposes of environmentalist theorising. On the contrary, it has deep and respectable philosophical roots. A form of an environmentalist account of the mind can be found in the work of Heidegger and also, to an extent, in Wittgenstein (when he wasn't in neo-Kantian mood). In recent analytical philosophy, one of the principal steps in this direction was taken by Hilary Putnam, Tyler Burge and others, with their development of a view often known as *externalism*.[14] Meanings, as Putnam demonstrated, 'ain't in the head'. Neither, as Burge and others have pointed out, are beliefs, desires, hopes, fears, or propositional attitudes in general. Rather, if we hold constant what is going on in the heads of thinkers, and vary their environment, their thoughts can change with changes in the environment, even though what is going on in their heads remains the same. Some have interpreted Putnam's and Burge's arguments as showing that environmental items can actually be constituents of thoughts, an idea with obvious environmental affinities. My development of the environmentalist model of cognition, however, rests on somewhat different considerations, from considerations pertaining to evolutionary theory, to the work of various psychologists including, notably, J. J. Gibson, and from certain very recent work in robotics and artificial intelligence.[15] The eclecticism of the supporting considerations, then, is hopefully by itself sufficient to undermine any suggestion that this is an *ad hoc* model of the mind constructed expressly for environmental purposes.

Furthermore, the development of this environmentalist model of the mind is not, in any way, intended as a *refutation* of neo-Kantian idealism. The environmentalist model, for example, claims that mental processes such as perceiving, remembering, thinking and the like contain certain environmental structures as constituents. But such structures are precisely the sort of thing that the neo-Kantian will claim are constituted by the structuring activities of the mind. Hence, the environmentalist model would backfire if presented as a refutation of neo-Kantianism. It is, in fact, very difficult to refute neo-Kantian idealism. The only method, I think, which has the remotest chance of succeeding is indirect, and involves tackling individual articulations of the neo-Kantian picture in a piecemeal manner and showing that the arguments used to support these articulations are defective. Such a project, however, is not undertaken here.[16] The near immunity to refutation possessed by the neo-Kantian picture is not, however, a sign of its coherence or power, but of its lack of empirical content. Neo-Kantian humanism is, ultimately, not a theory of the nature of reality but a pretheoretical picture that organises our thinking about reality. The force of this, and other pictures like it, is that they tempt us into thinking that *this is the way things must be.* And the role of the philosopher when presented with such a picture is not to refute it, but to undermine it; to show that things do not, in fact, have to be this way at all. This is the task of this book, and this is the purpose of the environmentalist model of the mind developed in later chapters. Ultimately, we have such difficulties in developing a satisfactory account of the mind because we are in the grip of a pre-theoretical picture, a picture with its roots in Descartes and, later, in Kant. The picture constrains our thinking about environmental value, and the task of this book is to remove these constraints.

1.4   The danger and that which saves

In his famous essay on the environmental crisis, 'The Question Concerning Technology', Heidegger quotes the poet Hölderlin: 'But where the danger is, grows also that which saves.' This, in many ways, could be taken to be the defining epigram of the present book.

In order to develop an adequate account of the value of nature, we need to break down the subject/object distinction. And we need to do this in a way that is not humanist, but rather which acknowledges the reality, indeed primacy, of the environment. We can never develop an adequate account of the value of nature if we insist on relegating nature to a secondary or derivative status, as ontologically and epistemologically subordinate to the structuring activities of the mind. Unfortunately, we have grown up in a neo-Kantian philosophical milieu that denigrates the natural environment in precisely this way. The way to break down the subject/object distinction in a way that is conducive to thinking that is genuinely environmental is to develop an environ-mentalist model of the mental, a model according to which mental operations are themselves worldly or environment-involving. In order, that is, to understand the nature of environmental value, we need to understand the environmental nature of valuing. This understanding breaks down the subject/object distinction by pulling the mind out into the world, not pulling the world into the mind. However, I shall argue that the reason why mental processes such as thinking, remember-ing, reasoning and perceiving are worldly, the reason why they have environmental structures as constituents, is, to a considerable extent, because these processes have developed in conjunction with abilities to manipulate and exploit environmental structures. Our manipulation and exploitation of the world is, in other words, essential to the proper collapse of the subject/object distinction, and the subsequent develop-ment of a realistic account of the value of nature.

The danger, a danger that has arguably precipitated us into the sort of environmental crisis we face today is our manipulative and exploitative nature. This is not a superficial part of us. It is not a peculiarly cultural product. Neither is it a product of gender. Rather, it has roots that delve deep into our natural (i.e. biological) history. It is our manipulative and exploitative nature that makes us like a cancer on the environment. However, ironically but not, I think, paradoxically, it is also this nature which makes us one with the environment; it is this nature which makes us not individual, isolated, monadic entities but, rather, genuine beings-in-the-world. This is what makes us zooids in *phrysalia*. Where the danger is, there too grows that which saves. Or, at least, so I shall argue.

2
Intrinsic Value and Why (We Think) It's Needed

Among the less desirable results of human activity on this planet are the following: pollution of atmosphere, rivers and oceans, land degradation, deforestation, elimination of species at an unprecedented rate, the build up of greenhouse gases, the depletion of the ozone layer. This list could easily be extended. It is generally assumed that the results are undesirable, that they result in the destruction of something with positive value. But what? What is the value of the natural environment that would be lost if the environment were destroyed or seriously degraded? This, essentially, is the central question of environmental ethics.

We might usefully organise our thinking on this matter by way of the following scenario.[1] Suppose Robinson Crusoe, upon completion of the building of the pinnace that was to take him off his island, decides to completely destroy his former home. He sets about laying waste to the island, slashing and burning his way across it, until nothing is left but a lifeless desert. Most people, I think, would be willing to accept that Crusoe has done something wrong here; that he has destroyed something of value. But what? What sort of value does the island possess, and why does it possess this value?

2.1 Human based value

At the very least, what is destroyed by Crusoe's slash and burn spree might have utility for human beings. No one was on the island during Crusoe's destruction of it. But, let us suppose, the island was regularly visited by Friday's people, who regarded it as valuable in various ways: as a useful repository of resources such as food and water, as providing certain useful herbs necessary for their medical practices, as a convenient place for rest and recuperation after a tiring sea voyage, and so on.

Similarly, what is destroyed by environmental degradation can possess an *instrumental value* for human beings. Since we depend, at least to some extent, on the integrity of the natural environment, the destruction of this environment goes against our interests. Human modifications of the environment, then, can clearly have a negative impact on human interests. Opinions are divided over the extent to which this is so. Some claim that such developments threaten the very continuation of human civilisation, making our continued existence tenuous.[2] Others, although not envisaging such dramatic consequences, nevertheless warn of the adverse impact of such developments on human beings, and the loss of scientific, medical, economic, aesthetic and recreational resources that such developments occasion. Furthermore, it might also be argued that human welfare can be dependent on the integrity of the natural environment in less obvious ways: for example, that human beings are such that they can properly flourish only if they have frequent contact with wild nature.[3]

It is fairly clear, then, that prudence, if nothing else, should make us concerned about the fate of the natural environment. At the very least, the natural environment has an instrumental value for us. It has value because it plays a role in supporting and furthering human interests. According to what we can call *human based approaches* to environmental value, the value of the environment consists *only* in this. That is, the value of the natural environment consists solely in the role this environment can play in supporting, underwriting and assisting in the satisfaction of human interests. The value of the environment, ultimately, consists in what it can do for *us*. Correlatively, the harm done to the environment by the effects of human activity consists solely in the negative impact these have on the furthering of human interests.

If we assume that the value of the environment consists solely in its instrumental value for human beings, then it is possible to apply orthodox ethical theories to environmental problems, and from them derive quite far-ranging environmental policies. An orthodox ethical theory, in this sense, is one that has been developed specifically to adjudicate relations between human beings. For example, according to the ethical theory known as *utilitarianism*, morally correct actions or policies are those which produce the greatest amount of happiness in the world, or which satisfy the greatest number of preferences.[4] If we accept that human happiness, or the satisfaction of human preferences, depends, at least in part, on there being a healthy natural environment, then utilitarianism would seem to be capable of underwriting substantive policies of environmental protection. Indeed, for utilitarianism, what

is important is the production of happiness or the satisfaction of preferences. Whose happiness is produced, or whose preferences are satisfied is not important. More importantly, *when* this happiness is
produced or these preferences are satisfied is also not directly relevant.
What is crucial is that the maximum amount of happiness is produced,
or the maximum number of preferences satisfied. When this occurs is,
for the utilitarian, morally irrelevant. Therefore, the utilitarian account
of morality is arguably capable of underwriting a moral commitment to
future generations of humans. If we assume, as we surely must, that
environmental modifications of today can have a serious negative
impact on the happiness or the satisfaction of preferences of future
generations, then this gives us a reason, a utilitarian reason, to act so
as to prevent, or at least inhibit, such modifications.[5]

Nor is this concern for future generations restricted to utilitarianism.
One of the major theoretical alternatives to utilitarianism is known as
contractarianism or *contractualism*, developed, in its most influential
recent form, by John Rawls.[6] Rawls asks us to imagine a situation
which he calls the *original position*. In this position, we are to imagine
that we have no specific knowledge about ourselves. We don't know if
we are male or female, black or white, rich or poor, intelligent or stupid,
and so on. Then, on the basis of this ignorance, we are asked to choose
the moral rules we would like to see adopted in the society in which we
are going to live. The principle is very much like being asked to divide
up a cake without being told which slice you are going to be given. The
rational thing to do in these circumstances, it seems, would be to divide
the cake evenly. Your ignorance of which slice you are to be given
removes any basis for partiality in your slicing. Your division of the
cake will, therefore, be a fair one. Similarly, on Rawls' view, in the
original position your ignorance of who you are removes any basis for
partiality in the moral rules you would choose for your society. The
rules you choose will, therefore, be fair ones. Now, one of the items of
knowledge that can, arguably, be excluded in the original position
is knowledge of *when* it is you are going to live, of which generation
you are going to be born into. Therefore, Rawls' account might underwrite a concern for future generations just as easily as utilitarianism.
Thus, Rawls' version of contractarianism can, arguably, also legitimise
the adoption of a wide-ranging set of environmental policies.[7]

It would be misleading to suggest that the application of either utilitarianism or contractarianism to future generations is uncontentious.
On the contrary, there are well-known problems with the attempt to
apply orthodox ethical theories to persons who do not yet exist, and

some of these problems are quite difficult. Nevertheless, it seems, at least in principle, that even if we adopt a human-centred approach to environmental value, and assume that the value of the environment consists solely in its capacity to further human interests, we might still be able to justify substantial policies of environmental protection, protection which is necessary to safeguard the well being of not only this generation of humans, but also future generations. Given the evident ethical scope of a human-centred approach to the environment, the question now naturally arises of whether we need anything more. Indeed, some have argued that we do not. Bryan Norton, for example, argues that a suitably subtle and extended human-centred approach will converge on precisely the same environmental policies as an approach based on what is known as the *intrinsic* value of nature.[8] We must remember that human interests in the environment are diverse and manifold, and cannot be reduced to the nakedly fiscal or the crudely exploitative. Once we take the whole spectrum of human interests into account, both short and long term, this will yield all the environmental protection we need. For this reason, Norton regards the debate over whether nature has intrinsic or merely human based value as an unnecessary split in the environmentalist camp, a split which hinders its effectiveness in the face of the forces of environmental destruction.

I think, however, that Norton's claim is probably false. The human-centred approach, it is argued, does not, despite its initial promise, provide an adequate moral basis for our dealings with the environment. To begin with, it is not clear that the human-centred approach is sufficient to justify the ecological integrity of the environment. Humans can apparently live quite comfortably in a world which contains monoculture tree farms instead of multiculture old forests. Humans have no need of mosquito infested swamps, especially when the land could be drained and used for a housing development. Humans could live, and apparently live well, in a world where most of the pristine stretches of wilderness had been transformed into cultivated fields and parks. Any part of the environment that is not required for the furthering of human interests cannot be legitimately protected on a human-centred approach. It is true, of course, as Norton points out, that we have many different sorts of interests in the environment, some of which may not be promoted by monoculture tree farms, housing developments, and cultivated parks and fields. But conflict of interests is part of the human condition, and often we are prepared to let some of our interests – those we judge to be of least importance – be simply overridden in order to satisfy more pressing ones. Now consider a part of the environment that is not instrumentally valuable to humans in any of the more obvious ways, say a stretch of wilderness. Here, the danger is that the sort of interests that are most likely to underwrite wide-ranging and long-term policies of environmental protection – perhaps recreational, aesthetic, religious, or more broadly spiritual – are precisely those interests of which we are likely to let go in the face of driving economic necessity (or what is perceived as such). So, it is not at all clear that a suitably subtle, wide-ranging and long-term human based approach will underwrite sufficient protection for all parts of the natural environment.

Secondly, even if this problem can be overcome, it is still true that those parts of the environment that are, in fact, judged to be essential to human interests can receive only *conditional* protection on the human based approach: their future protection is conditional upon them continuing to be necessary to further human interests. The presence of rain forests, for example, is considered by most to be required for the promotion of human interests due to the significant role they play in stabilising the Earth's climate, among other things. But suppose that, some time in the future, technology advances to a level where we can build devices that simulate the role of rain forests in world climate regulation. Indeed, suppose that at some time in the future it was possible to build devices that replicate whatever beneficial effects – climatic, atmospheric, scientific, medical and so on – that rain forests have. Under these circumstances, rain forests would no longer be required to promote human interests, and we could, on a human centred approach, legitimately eradicate them from the face of the Earth. Similar points apply to any environmental feature whose beneficial effects can be simulated by a device of human construction. And one corollary of this seems to be that the more that technology advances in the relevant way, the more of nature we can legitimately destroy. And this is a claim that, presumably, not many environmentalists would be prepared to endorse.

Third, and perhaps most importantly, human centred approaches are at serious *strategic* disadvantage compared to an approach which claims that the environment has more than a merely human based value, that it is *intrinsically* valuable. The claim that nature has intrinsic value – that it is valuable in and of itself – if true, puts the *burden of proof* on those who would interfere with, manipulate or exploit it. But the claim that nature has only instrumental value puts the burden of proof on those who would prevent such interference. This point has been well made by Warwick Fox.[9] As Fox points out, to claim that something is intrinsically valuable does not mean that it can never be interfered with under any circumstances. Human beings are generally accepted as having intrinsic value, but in certain circumstances it is morally legitimate to imprison, put at risk, perhaps even kill, some of them. But, if we are to justify such treatment, sufficient justification must be given. Similarly, Fox argues, in the case of the environment. If the environment has only instrumental, or human based, value, then people are morally permitted to interfere, manipulate or exploit it for whatever reason they wish, i.e. no justification need be given. And if anyone objects to such interference, then the burden of proof is on them to justify their objection: that is, they must demonstrate why it is more useful to human beings to refrain from interfering with that part of the environment. However, if the environment has intrinsic value, then the burden of proof swings to the person who would interfere with it. Now it is *they* who must provide sufficient justification for what they propose to do. This, as Fox points out, represents a fundamental shift in the terms of environmental debate and decision-making. Thus, *pace* Norton, the question of the status of the value of the environment does make a huge practical difference.

Therefore, in the eyes of many environmentalists at least, it would be desirable if the environment were to possess more than simply human based, or instrumental, value. This, of course, is no reason for thinking that it does possess this additional value. Wishing does not make something so. And whether or not the environment does possess a value that is not human based will be the subject of the coming chapters. But this attitude does explain the search by philosophically oriented environmentalists for value in nature over and above its instrumental value for human beings. The question, then, is what sort of value this might be?

2.2   Sentience based value

In slashing and burning his island, Crusoe destroys more than just a human resource. He destroys the homes, livelihoods and in all probability the lives of the creatures that reside there. Moreover, many of those creatures will be conscious or sentient, and Crusoe's actions would lead to an enormous amount of suffering for these creatures. Therefore, it seems that, at the very least, the island would have value not only for the role it plays in furthering human interests, but also in furthering the interests of the sentient creatures that live there.

Most of us, if we are not psychopaths, allow that our treatment of sentient non-human animals does raise moral issues of some sort. While most people would agree that, special circumstances aside, it is morally acceptable to take a chainsaw to a living tree, those same people would be horrified if we took a chainsaw to a living cow. And rightly so. The crucial difference, at least in the eyes of many, is that the cow is, whereas the tree is not, sentient. This is why, of course, we have various laws prohibiting cruelty to animals but not to trees. At the same time, many people are unwilling to allow that animals are, so to speak, full members of the moral club. We treat them in ways we would not, or should not, dream of treating humans. We eat them, we experiment on them, we hunt them, and we wear them. What can justify this difference in the treatment we accord humans and non-humans?

In the eyes of many philosophers, *nothing at all* justifies this differential treatment. Our current treatment of animals is morally very wrong indeed. Many non-human animals possess a substantial set of moral entitlements, even though these are largely unacknowledged by the bulk of the human population. Peter Singer has argued, very plausibly, that utilitarianism underwrites the granting of many moral entitlements to sentient creatures, including, principally, the negative entitlement to not have unnecessary suffering inflicted upon them.[10] Tom Regan has made an equally compelling case for the claim that many non-human animals possess moral *rights*.[11] And, elsewhere, I have argued that the most plausible versions of contractarianism entail that many non-human animals possess a substantial set of moral entitlements.[12]

Despite the diversity of moral theories involved, what underlies these, and indeed most, arguments for the moral claims of animals is a line of thought that can be developed in the following way. Moral thinking is governed by a principle which can be stated, roughly, as follows: *no moral difference without a relevant other difference*. For example, Hitler is an archetypal example of someone who is morally very bad. What made him bad? Well, different ethical traditions provide different answers to this question. Ethical theories that are consequentialist, and here utilitarianism provides the most obvious example, emphasise the consequences of people's actions. The rightness or wrongness of an action turns solely on its consequences. Other views, known as deontological moral theories, emphasise the intentions with which actions are performed. The rightness or wrongness of an action depends, at least in part, on the intentions with which the person acts. But, to avoid adjudicating between consequentialist and deontological positions, let us suppose that there is a person who does the same things as Hitler; he performs the same actions with the same consequences and with the same intentions. He starts a world war, sends six million people to the gas chambers, and so on, and does so for exactly the same reasons as Hitler. In this case, it would *make no sense* to say that Hitler was a bad person but this other person was not. If Hitler is bad, and if this other person does the same things as Hitler for the same reasons, then this other person *must* be bad also. A difference in moral evaluation only makes sense if it is based on a difference in other qualities. In fact, not only must there be a difference in other qualities, this difference must also be a relevant one. Thus, if this other person carried out his atrocities in, say, east Asia instead of Europe, this would not, presumably, be a relevant difference. The same sort of point applies not just to the moral evaluation of persons, but to anything that can be morally judged or evaluated. Any difference in moral evaluation of two items – persons, actions, rules, attitudes, institutions – must be grounded in and reflect a relevant difference in other (i.e. non-moral) qualities possessed by those items.

The moral is that if we are to justify the differential treatment we accord human and non-human animals, we must be able to cite a relevant difference between the two, and show how the differential treatment can be justified on the basis of this difference. There are many differences between humans and non-humans that might suggest themselves as morally relevant ones: intelligence, rationality, self-consciousness, capacity to conceive of the future, linguistic abilities have all, at various times and by various people, been suggested as decisive. However, these suggestions all seem to face an apparently fatal objection. Even if we allow that all non-human animals do not possess these features (and, for some of the features this seems very questionable), it is also true that many human beings do not possess these features either. Human beings who do not seem to possess *any* of these features include the severely brain damaged, infants, and those in advanced stages of senility. And human beings who do not possess at least some of these features include the moderately brain damaged, young children, the permanently insane, the temporarily insane, and, perhaps, the moderately senile. Thus, if we are prepared to withhold moral status from non-human animals because of their lack, or alleged lack, of these features, then we must be prepared to withhold moral status from the relevant categories of human beings. If there is no moral obstacle to our eating, vivisecting, wearing and hunting non-human animals, then equally there can be no moral objection to our doing the same to the relevant categories of human beings.

Of course, most of us find the idea of eating or vivisecting infants abhorrent. Nor are we impressed with the idea of saddling up the horses, releasing the hounds, and chasing the senile all over the countryside. And we would not dream of wearing jackets made from the skin of the brain damaged or insane. And this shows that our moral judgements are not, in fact, guided by the sorts of features listed above. And this shows that we do not regard the features as morally relevant ones. In fact, our treatment of humans from the various categories listed above seems to indicate that we regard sentience as the morally decisive factor. Our treatment of the brain damaged, infants, the insane and the senile is guided by the fact that they possess sentience, and not by the fact that they lack rationality, or are of lesser intelligence, or whatever. And this shows that, with respect to the possession of moral entitlements, we think of sentience as crucial. But many non-human animals are sentient. And this means that they must possess moral entitlements.

Therefore, Crusoe's destruction of his island destroys a thing that is of value to various creatures that reside there. And, it seems, at least some of those creatures – the sentient ones – possess moral entitlements. And this has suggested to many that our environmental decision-making should be guided not just by considerations of instrumental value of the environment to human beings, but also of its instrumental value to all sentient creatures. This move is one from a human centred approach to the value of the environment to a sentience, or consciousness, centred approach. The value of the environment consists in its instrumental use for all sentient creatures.

2.3   Environment based value

In adopting the sentience based approach to environmental value, we have at our disposal the same meta-ethical resources as the human centred approach. This is because the classical ethical theories such as utilitarianism, contractarianism, natural rights theory and so on were designed to deal with human interactions in virtue of the fact that humans possessed such features as needs, interests and desires, were capable of feeling pleasure and pain, and so on. Utilitarianism, for example, states that we should increase pleasure and/or decrease pain, or that we should maximise the amount of satisfied desires. The argument for incorporating at least some animals within the scope of morality was based on the claim that they too possessed these sorts of features. Therefore, it is a relatively small step to modify our human based ethical theories to incorporate non-human animals. Indeed, many have argued that such an extension was always implicitly mandated by the very content of these theories.[13]

A consciousness based approach to environmental value will, presumably, significantly increase the scope and requirements of environmental policy. Now, in our environmental decision-making, we have to protect the interests not just of human beings, but of all sentient creatures. The mosquito infested, leech ridden, moccasined swamp might not further any human interests, but its preservation might be in the interests of the sentient creatures residing there. Therefore, a consciousness centred approach would seem to afford a significant increase in environmental protection. It, therefore, would be welcomed by many environmentalists.

However, a sentience based approach does not seem adequate to the requirements of many environmentalists.[14] This is so for two reasons. Firstly, many would want to claim that the environment possesses a value over and above its being valued by sentient creatures, human or otherwise. Suppose Crusoe's island were not inhabited by any sentient creatures, and that the prevailing winds and currents coupled with its remoteness made any habitation by such creatures in the future practically impossible. Nevertheless, the island constituted a complex ecosystem, made up of an intricate web of flora and non-sentient fauna. Rare lichens intermingled with breathtakingly beautiful butterflies in a delicate yet profoundly balanced dance of life. You get the picture. If we adopt a sentience based account of environmental value, then in turning the island into a lifeless desert Crusoe *has done nothing wrong*. There are no human interests left to further since Crusoe is leaving the island, and there are no wider sentient interests to further since the island is inhabited by no sentient creatures. But to many environmentalists, this seems, to say the least, counterintuitive. When Crusoe turns the island into a lifeless desert, he certainly seems to be destroying something of value. And if this intuition is correct, then the value of the environment cannot be reduced to the role it plays in furthering the interests of sentient creatures, human or otherwise.

Secondly, there are situations in which the environmental policy that would seem to be mandated by a sentience based approach can conflict with that recommended by many environmentalists. Suppose Crusoe managed to offload some sheep from the ship before she went down. Upon reaching the island, the sheep proceed to multiply rapidly, and after a number of generations, they are sufficiently numerous to wreak havoc on the island's ecosystem. To be sure, they do not destroy any sentient life (since, besides Crusoe, they are the only such life on the island), specialising instead, let us suppose, in a certain type of plant which is extremely rare, indeed which happens to be extant only on that island. The sentience based criterion of environmental value, it seems, would allow the entitlements of the sheep to override any consideration of the integrity of the island's ecosystem or of the rarity of the plants they eat. They are sentient, and, therefore, whatever value the island has derives from them (and, of course, Crusoe; but let us suppose that the sheep's activities in no way impinge on his interests). But this flies in the face of the attitudes of many environmentalists who would advocate a widespread culling of the sheep in order to safeguard the island. The situation is a common one, perhaps the most leading cause of contention between environmentalists and animal rights activists. And the attitude of the environmentalists, here, would seem to indicate that they believe the island's environment has a value over and above its role in furthering the interests of sentient creatures.

This point can be pushed further by considering the difference between the attitude of an environmentalist and that of an animal rights activist towards the following choice. You have a choice of saving the life of two creatures. One is the sentient member of a widespread species, one of Crusoe's sheep, for example. The other is one of the last surviving members of a non-sentient species; an endangered species of plant that Crusoe's sheep is about to eat. The only way of saving the plant is by shooting the sheep. What should you do? An animal rights activist, motivated by the sentience based criterion of environmental value, would say that you should allow the sheep to eat the plant. The environmentalist would tell you to shoot the sheep and save the plant. If we base our moral decision-making purely on a sentience centred approach, then the fact that an individual animal is among the last remaining members of a species confers no further value, hence no further entitlements, on that animal. Thus, if we had to choose between saving a member of an endangered species and a member of a species that was plentiful, then our decision should turn only on the issue of sentience. If the member of the endangered species is non-sentient, but the member of the plentiful species is sentient, then we should save the latter rather than the former.[15] This contrasts with the attitudes of most environmentalists who regard a property such as rarity as conferring a special value or status on whatever individuals possess it. And, once again, the environmentalist's attitude indicates that, for her, the value of nature cannot be reduced to its role in furthering the interests of sentient creatures. For her, the environment has value in addition to its role in furthering the interests of sentient creatures. It has what is sometimes called *intrinsic value*.

The position, then, is this. At least some of the value the natural environment possesses arises because of its capacity to satisfy or promote human interests. Let us refer to this as the *human based value* of the environment, indicating that it is value which has its origin or source in human valuing of the environment, where this valuing arises because of what the environment affords humans, whether this be economic, scientific, medical, aesthetic, recreational, spiritual, or whatever, in character. In addition, some of the value the natural environment possesses derives from its capacity to satisfy or promote the interests of sentient creatures in general, whether human or otherwise. Let us refer to this as the *sentience based value* of the environment, indicating that this is value that has its origin in the valuing of the environment by sentient creatures. Human based value is, thus, a subset of sentience based value. However, many environmentalists would go further and claim that the environment has value over and above both human based and sentience based value. The environment also has *intrinsic value*.

At least as a first approximation, then, we can say that the claim that the environment has intrinsic value is the claim that not all the value possessed by the natural environment consists in human or sentience based value. As we shall see, this claim is not as straightforward as it seems, and the line between value that is human or sentience based and value that is not is not as distinct as it might first appear. However, this gives us a place from where to start.

Given this starting point, we can say the following. The possibilities of developing a genuine environment based account of the value of the environment, depend on us making sense of the idea that the environment possesses a value over and above its human or sentience based value. It requires, that is, that we make sense of the idea that the environment possesses what is often referred to as *intrinsic* value. If we cannot make sense of this idea, or of the notion of intrinsic value which underlies it, then there seems to be no room for an environment based environmental ethic. If environmental value reduced to human based value, then environmental ethics would be restricted to a human centred approach. It would be restricted, that is, to developing a prudential account of how human beings can best use the environment to further their own purposes. If environmental value reduced to sentience based value, then environmental ethics would effectively collapse into a case for the moral entitlements of sentient creatures, coupled with an account of how the environment is best used to further the interests of such creatures. Neither of these projects, however, provides a genuine ethics *of* the environment. At most they provide an ethics *for* environmental use.[16] That is, they do not provide an account of how our treatment of the environment should be constrained by, and reflect, the value that it possesses in and of itself. Rather they would provide an account simply of how the environment is to be best used to further the interests, needs and purposes of human and other sentient creatures. For a genuine ethics *of* the environment, it seems, we need the claim that the environment has *intrinsic value*. But what does this claim mean?

2.4 Concepts of intrinsic value

In elucidating the idea of intrinsic value, the first problem we face is that the concept of intrinsic value, as it appears in the environmental ethics literature, is not univocal. In fact, it is possible to discern at least three distinct concepts of intrinsic value. Sorting through these distinct concepts, and working out which are relevant to the attempt to develop an environment based ethics is our first task.

The first sense of intrinsic value, which we may call *intrinsic value₁*, identifies intrinsic value with the value that an object possesses in virtue of its intrinsic properties or features. This was the sense of intrinsic value employed by one of the great intellectual progenitors of that concept, G. E. Moore: 'To say a kind of value is "intrinsic" merely means that the question of whether a thing possesses it, and in what degree it possesses it, depends solely on the intrinsic nature of the thing in question.'[17] A value is intrinsic when its possession by an object depends only on the intrinsic properties of that object. And a property is said to be intrinsic to an object if it is a *non-relational* property of that object. And a property is non-relational if it satisfies one or the other of the following conditions: (i) the possession of the property by the object does not depend on the existence or non-existence of other objects, or (ii) the property can be identified without reference to other objects. The difference between (i) and (ii), while important in some contexts, is, however, irrelevant to our purposes, and I propose henceforth to gloss over it.

The concept of intrinsic value in this sense, however, does not seem to be the one required for the purposes of an environment centred approach to environmental value. The reason is that some of the relational properties of environmental items seem to be ones that contribute to its intrinsic value. *Rarity* provides a good example here. The property of being rare is an irreducibly relational one, in both senses described above. Whether or not an object is rare depends on the existence or non-existence of other objects, and the property cannot be identified or characterised without reference to other objects. Thus, it is not an intrinsic property of objects. Therefore, if the intrinsic value of an object were to depend only on its intrinsic properties, we would have to say that being rare confers no value on an object. But, as we saw in the case of Crusoe's sheep, this contradicts a central strand of environmentalist thought which regards the property of being rare as conferring value on objects which possess it. The preservation of endangered species of flora and fauna is a major practical environmental concern. Similarly, many environmentalists argue that *biodiversity* is a fundamentally valuable feature of the environment, and that attempts should be made to protect and promote biodiversity in ecological systems. And one of the constituent concepts here – the concept of diversity – is also irreducibly relational. Finally, being naturally evolved, as opposed to artificial, is a property which, for most environmentalists, confers value on those things that have it. But the property of being naturally evolved is a historical property: whether or not an object has it depends on certain facts about its history. And historical properties are also essentially relational properties, it is just that they are made up of relations to the past rather than to presently existing entities. In short, to understand intrinsic value as value determined by the intrinsic properties of objects is inadequate for the purposes of an environment based ethic. Many, and perhaps most, of the properties that intuitively seem to confer intrinsic value on objects are irreducibly relational properties of those objects.

We will more closely approximate the conception of value required for environmental purposes if we, first, distinguish two types of relationally constituted value that objects might possess. On the one hand, there are values that objects can have in virtue of their relations to other objects. On the other hand, there are values objects can have in virtue of the relations to human beings or other sentient creatures. What is required for the purposes of an environment based ethic is the exclusion of the latter sort of value; value that objects have in virtue of their relations to human beings or sentient creatures in general. It does not require, and indeed cannot accommodate, exclusion of values of the former sort; values that objects possess in virtue of their relation to other objects. However, if we exclude only the latter sort of value, then our interpretation of the notion of intrinsic value falls into one or other of the remaining two construals of that notion. It is to these that we now turn.

According to the second interpretation of the concept of intrinsic value, which we shall refer to as *intrinsic value$_2$*, this type of value should be understood as *non-instrumental* value. Essentially, an object has instrumental value in so far as it is a means to some other end. And, on this construal, an object has intrinsic value in so far as it is an end in itself. An intrinsic good is a good that other goods are good for the sake of. It is widely thought that, under pain of infinite regress, not everything can have merely instrumental value. There must be some objects that have intrinsic – in the sense of non-instrumental – value. And some defenders of an environment based ethic have claimed that among these objects that have non-instrumental value are environmental objects, states and processes.

The third interpretation of the notion of intrinsic value, *intrinsic value$_3$*, understands this as value that is *objective* in the sense of not being dependent for its existence on the opinions, feelings or attitudes of sentient creatures. Intrinsic value, in this sense, attaches to objects irrespective of whether they are valued by sentient creatures. While some of the value a natural object possesses might be value that is projected upon it by sentient valuers, the claim that the object has intrinsic value is the claim that not all of its value is like this. Some of the value it possesses it does so independently of sentient attitudes towards it.

It is important to realise that intrinsic value$_2$ and intrinsic value$_3$ are not equivalent, although they are often conflated. Firstly, to claim that the value of an object is instrumental value is not to claim that it is subjective. The instrumental value of a heart, for creatures that have one, is that it pumps blood. But this is a perfectly objective claim in the sense that its truth does not depend on the feelings, opinions or other attitudes that we might bear towards hearts. The truth of this claim about the value of a heart would not change, for example, if everyone were ignorant of it. Or if nobody believed it. The fact that an object has a certain instrumental value can be a perfectly objective fact about the world. Therefore, instrumentality does not entail subjectivity. Secondly, to claim that a value is subjective does not entail that the value is instrumental. Many things can be subjective, or subjectively constituted, without being instrumental. The way something tastes, the way something feels, or, more generally, what are known as the *qualia* of experience, are all subjectively constituted, but not all of them are instrumental. Therefore, subjectivity does not entail instrumentality.

If we accept, as it seems we must, that intrinsic value$_2$ and intrinsic value$_3$ are distinct concepts, the question now arises of which one is required for the purposes of developing an environment based ethic.

This question has no straightforward answer, and attempts to provide answers will, in effect, be the subject of Chapters 3 and 4. However, at this point we can give some idea of the issues involved, and of the form these answers might take.

Firstly, consider intrinsic value$_2$, intrinsic value as non-instrumental value. Adopting this conception of intrinsic value seems an absolute requirement of the attempt to develop an environment centred ethic. If the only value nature has is instrumental in character, then no matter how objective this instrumental value is, the value of nature will depend solely on the role it plays in furthering human or sentient interests. And this means that any attempt to develop an environment based ethic will quickly collapse back into either a human based prudential ethic or a case for animal rights. If the value of the natural environment lies solely in the role it can play in promoting human, or more generally sentient, interests, then the attempt to develop an environment centred ethic is redundant: all substantive moral issues concerning the environment can be dealt with in terms of a *shallow* ecological ethic, an ethic of environmental management, either for the benefit of humans or for that of all sentient creatures. Therefore, the claim that at least some of the value of nature is non-instrumental seems to be essential to an environment based ethics.

However, while the claim that at least some of the value of the natural environment is not instrumental in character seems to be *necessary* for a deep ecological ethic, it does not, at least at first glance, seem to be *sufficient*. For if we were to claim that the value of nature, while non-instrumental, is nevertheless subjective, then it seems, at least at first glance, we would also very quickly be led back to a sentience based ethic. This is so because to claim that the value of nature is subjective is to claim that it depends for its existence on the attitudes that sentient creatures bear towards it. For example, it is to claim that it depends for its existence of the fact that sentient creatures value it; or on the fact that such creatures in some way approve of it. Generally, to say that value is subjective is to say that it exists only because of, and only in relation to, mental states such as valuing, approving, and the like. But this claim seems to entail that the question of how, morally, we should treat the environment is to be answered in accordance with the dictates of these mental states. Features of the environment towards which we, or other sentient creatures, are favourably disposed (i.e. we value them, or approve of them) should be protected. Features towards which we disapprove, or towards which we are indifferent, should not, morally speaking, be protected. But once again, this leaves us with a sentience based ethic: how we should treat the environment depends on, and only upon, the attitudes of sentient creatures towards it. The environment has no value other than what is invested in it by the mental states of human or other sentient creatures. Therefore, it seems that the project of developing a genuine environment centred ethic requires the claim that nature has a value that is both non-instrumental *and* objective. That is, the project of developing a genuine ethic of the environment, as opposed to an ethic for how the environment should be used for the benefit of sentient creatures, seems to require the claim that intrinsic value is objective in the sense, roughly, of not being dependent for its existence on the mental properties of sentient beings. The project, that is, seems to require an *objectivist* account of intrinsic value.[18]

This claim, however, has been contested. Some have argued that the core of an genuine environment centred ethic is non-instrumentality rather than objectivity. More specifically, it is possible to develop a *subjectivist* account of intrinsic value that is adequate for the purposes of an environment centred ethic.[19] A subjectivist account of intrinsic value can take a variety of forms, but what is common to these forms is the idea that intrinsic value is dependent for its existence on the mental properties of either human beings or sentient creatures in general. This would seem to lead to the collapse of environment based value back into human or sentience based value. However, proponents of a subjectivist account of value deny that this is so. Central to the idea of a subjectivist account of intrinsic value is the distinction between the *origin* of such value and its *content*. This will be dealt with at some length later. For now it is perhaps sufficient to say that, according to proponents of the subjectivist account of intrinsic value, it is possible to construct an account of environmental value which, although having a subjective source in the valuings of human beings or sentient creatures, nonetheless has a content that is sufficiently objective, and sufficiently non-instrumental, to avoid the collapse of an environment centred ethic into a human or sentience based one. Subjectivist accounts of intrinsic value will be the subject of Chapter 4. The next chapter, however, is concerned with objectivist accounts of the value of the environment.

3
Objectivist Theories of Intrinsic Value

The preliminary arguments outlined towards the end of the previous chapter suggest that the possibility of developing a genuine ethic *of* the environment, as opposed to a human or sentience based ethic *for* environmental use, requires that the environment possess value that is both non-instrumental, in that it does not depend on its utility for humans, and also objective, in that it does not depend on the beliefs, attitudes, opinions or feelings of human or other sentient valuers. Any account of environmental value that satisfies, or purports to satisfy, these conditions of non-instrumentality and objectivity I shall refer to as an *objectivist* theory. Objectivist theories of environmental value are the subject of this chapter.

3.1 Objectivism about value

Broadly speaking, objectivist theories of environmental value are thought to take two forms. What is common to both is the idea that the value of the environment is objective in that it exists independently of the opinions, feelings, beliefs or attitudes of people, or of any valuing consciousness in general. The environment is valuable, that is, even if no one, or no thing, happens to value it. Within this general framework, however, opinions diverge over the precise nature of this value. One form of objectivism identifies environmental value with some group of natural features that are present in the environment. This form we can call *naturalistic objectivism*. The other form identifies environmental value with a so-called non-natural feature of the environment, though perhaps one that is associated with certain natural environmental features. I shall refer to this sort of position as *non-naturalistic objectivism*.

A naturalistic theory of value is, roughly speaking, one which identifies moral value or worth with a natural feature of objects, persons, rules, institutions or, in the present case, environments. And to say that a feature is natural is, roughly, to say that is capable of empirical detection, either through direct observation, or through inference from what is directly observed. Hedonism, for example, qualifies as a naturalistic theory in this sense. According to hedonism, pleasure is valuable, and therefore an increase in pleasure is morally good, and a decrease morally bad. One can observe, introspectively, one's own pleasure, and one can infer its presence in others through their behaviour. Pleasure thus qualifies as a natural property. Similarly, according to Kant's moral theory, reason is the ultimate value, and all rational beings are, therefore, morally valuable. Naturalistic versions of environmental value, therefore, are based on the idea that the environment possesses a certain natural feature or features, and that the intrinsic value of the environment is to be identified with this feature or these features. The environment, therefore, has intrinsic value in virtue of having these features. However, consensus has not actually converged on what these features are. One quite common view is *biocentrism*, the view that life is the crucial feature that lends intrinsic value to those things that have it, and this position is represented in the work of Goodpaster, Taylor and some of the work of Rolston.[1] One of the problems with biocentrism is that it is unable to grant intrinsic value to environmental structures such as ecosystems, or to environmental kinds such as species. And it partly with this limitation in mind that Miller has developed the idea that the crucial feature that lends intrisic value to the environment is what he calls *richness*, where this seems to be a complex concept involving such components as complexity, diversity and, perhaps, stability.[2] Whatever feature these positions identify as lending the environment intrinsic value, they are all united by the view that the relevant feature is a natural one.

An alternative is to adopt a non-naturalistic version of objectivism. According to non-naturalistic objectivism, the value of nature is not to be identified with any natural features of the environment. Rather, goodness, or intrinsic value, is a primitive and irreducible property that is objective but non-natural. To say that this value is non-natural is to say that it is not empirically detectable, either through observation or inference on the basis of observation. And to say that this value is primitive and irreducible is to say that the nature of this value cannot be explained in any other terms. This, in effect, is what naturalistic versions of objectivism try to do. They try to explain the nature of value by appeal to other properties: pleasure, rationality or, in environmental contexts, life, richness and so on. Non-naturalistic versions of objectivism deny that this can be done. Value is simply a primitive and irreducible property of certain things, and we can say nothing more about its essential nature.

It might be thought that the apparently clear line between naturalistic and non-naturalistic versions of objectivism tends to blur somewhat on further analysis. Most environmental versions of naturalistic objectivism do not claim that value is literally identical with life, richness or whatever. They allow, that is, that it is possible for there to be things that are intrinsically valuable which do not possess these properties. This is because things other than the environment or living things might have intrinsic value. Therefore, the claim, typically, is that these properties are *sufficient* for the possession of intrinsic value, not that they are *necessary* for the possession of this value. But this means that intrinsic value cannot be *identified* with any of these properties. Rather it must be regarded as *supervenient* upon them. The idea of supervenience can be explained as follows. Colour supervenes on redness, in the sense that any property which is red is also coloured, but *not*, necessarily, vice versa. Similarly, so naturalistic versions of environmental objectivism claim, intrinsic value supervenes on properties such as richness, in that any environmental system that is rich is also valuable, but not, necessarily, vice versa. But if this is so, then the intrinsic value of the environment must be a property that is distinct from richness, complexity, integrity, diversity and the like. Colour is not the same property as redness, since some objects which are coloured are not red. Similarly, intrinsic value is not the same property as richness, life, complexity, integrity, diversity, etc. since some objects have intrinsic value but not any of these other properties. But, then, the question arises as to what sort of thing this property of being intrinsically valuable is. And, to some extent, we would seem to be pushed back towards the mysteries of non-naturalism. Indeed, the supervenience of the non-natural property of intrinsic value on natural properties is a standard feature of non-naturalistic objectivism. And this makes it unclear in what respect the views differ.

The way to preserve the distinction between naturalistic and non-naturalistic versions of objectivism is to regard both forms as concerned not with the question of in what intrinsic value generally consists, but rather of the following, more specific one. When, on any particular occasion, the environment possesses intrinsic value (or is thought to do so), then in virtue of what does it possess this value? The naturalistic objectivist says that this will be solely in virtue of certain natural properties, the non-naturalist denies this. And for the naturalist but not the non-naturalist, the *instantiation* of intrinsic value in the environment can be identified with the *instantiation* of a certain natural property or properties. Compare the case of colour. The property of being coloured is clearly not identical with the property of being red, since objects can possess the former property without possessing the latter. Nevertheless, when an object is red, and therefore also coloured, it would be very implausible to say that the object possesses both an instance of redness and, *in addition*, a distinct instance of colour. In this particular case, the instance of redness just is (i.e. identical with) the instance of colour. Similarly, the naturalistic objectivist will say that when, on any particular occasion, the environment possesses intrinsic value, this possession or instantiation of intrinsic value is identical with the instantiation or possession of a certain natural property or properties. The non-naturalistic objectivist will deny this. Thus both forms of objectivism can be understood as concerned with the question of the nature of instantiations of intrinsic value in the environment, rather than the more general question of the nature of intrinsic value itself. And, when understood in this way, there *is* a clear distinction between the two.

3.2   Problems with non-naturalistic objectivism

According to non-naturalistic objectivism, intrinsic value cannot be identified with some natural property or set of properties. Rather, intrinsic value is a primitive or irreducible property possessed by certain things. One cannot give a naturalistic account of the essence of this property, nor can one give a naturalistic explanation of why some things have this property while others do not. Some things just have this property of being intrinsically valuable, while others do not. This is all we can really say. This is, however, compatible with intrinsic value being supervenient upon, or associated with, certain natural properties. It's just that these natural properties do not constitute the nature of intrinsic value.

It is very difficult to find environmental philosophers willing to endorse a non-naturalistic account of environmental value. Of course, it would I suppose be ironic should the value of the natural environment turn out to be a non-natural property. But the real problems with a non-naturalistic approach lie not in irony but in intellectual bankruptcy. Perhaps most importantly, it reduces moral argument to a mystical form of intuitionism. Natural properties of objects may be recognised or discovered empirically or by reasoning based on experience. We know, for example, that an object is circular by looking at it, and we know (colloquially) that our car's fuel tank is empty by inference from other experiences (the gauge points to empty, the car will not start, etc.). The non-natural property of intrinsic value, however, cannot by definition be empirically detected or inferred from other experiences that we have. Indeed, this is the force of saying that it is a non-natural property. The non-natural property of intrinsic value can only be discovered by some sort of obscure faculty, through a capacity of *moral intuition*. We can know moral truths, on this view, not through our experience or through our reason, but through our faculty of moral intuition. However, such a faculty must clearly be unevenly distributed throughout the populace. If such a capacity for moral intuition were genuinely distributed throughout the population, distributed equally and evenly in everyone, then there would be as little controversy about the intrinsic value of the environment as there is about the colour of snow. However, which things have intrinsic value and which do not is not something that is generally agreed on, and this is true not just in environmental contexts, but in moral contexts generally. This, after all, is why there are moral disagreements. Therefore, one must regard the faculty of moral intuition as being unevenly distributed between people. Perhaps it is vested in only a few gifted moral seers. Or perhaps, though generally distributed, it varies wildly from one person to another. Either way, the possibility of moral persuasion based on rational discussion is aborted. One can only say things like: I 'see' (i.e. intuit) the intrinsic value of X, and if you do not you are morally blind. The result is simply undefended, entrenched opinions clothed, for the sake of respectability, in the guise of intuitions of primitive, irreducible, non-natural properties.

It is fairly clear that this conception of intrinsic value would be useless for the purposes of rational defence of the environment. While intuitions about the value of human beings, and how we should treat them, have a reasonably wide currency, so that one might be able to get away with claiming that one just intuits the intrinsic value of humans, this does not seem to be the case for our intuitions about the environment. Just try telling a pacific Northwest logger that you just 'see' or 'intuit' the intrinsic value of the spotted owl and see how far you get. Any basis for an environment centred ethic that is to be both reasonable and defensible must be found in other than non-naturalistic objectivism.

3.3 Problems with naturalistic objectivism

One principal advantage of a naturalistic form of objectivism is, of course, that it avoids the sort of mystery inherent in its non-naturalistic counterpart. However, there are still serious and well-known problems with the naturalistic position. And the most basic problem is that it falls foul of what is known as the *naturalistic fallacy*, and does so in a way that makes naturalistic accounts of environmental value appear *arbitrary*.

The expression 'naturalistic fallacy' was first coined by G. E. Moore, and denotes the (alleged) fallacy of assuming that ethical words can be defined by way of non-ethical ones.[3] Thus, when Bentham, for example, says that pleasure is good, he is offering, in effect, a definition of good (a moral word) in terms of pleasure (a non-moral word). Thus, he commits what Moore calls the naturalistic fallacy. The naturalistic fallacy, in fact, is not really a fallacy at all, at least not in the logical sense, since a fallacy in the true sense of the word can only be committed in moving from the premises of an argument to a conclusion that does not logically follow from them, and no such move is involved here. For our purposes, what the notion of a naturalistic fallacy highlights is not so much any logical flaw in naturalistic objectivism, but rather the arbitrariness of this position.

To see this, consider one example of this position, associated with Kenneth Goodpaster.[4] Goodpaster claims that life is intrinsically valuable and, therefore, that all living beings have moral worth. As Regan points out, however, identifying the class of individuals that have inherent value with the class of living beings is an arbitrary decision. The sceptic is entitled to ask: Why is life crucial? Why not some other feature, such as consciousness or rationality? A similar point applies to Peter Miller's attempt to define the intrinsic value of nature in terms of the notion of richness.[5] According to Miller, richness is the property which makes people, other organisms, and nature as a whole intrinsically valuable. Again, the sceptic is entitled to ask why richness should be accorded this crucial role. Why is richness so important? Apart from our valuing richness, or having some sort of preference for richness, why is richness important in any objective sense?

The problem is that one cannot simply assert that a certain property is what gives things which have it intrinsic value and leave matters at that. Some sort of defence needs to be given of why this property is so important, of why this property in particular makes something intrinsically valuable. The problem for naturalistic objectivists is that they seem, in principle, incapable of providing such a defence. For naturalistic objectivists, the explanatory buck stops with the natural properties themselves. There simply is no explanation of why life, or richness, makes something intrinsically valuable; it just does. Explanation comes to an end with the connection between these sorts of natural properties and intrinsic value. Now, there is no problem as such with the notion of explanation coming to an end. All explanation, as Wittgenstein has pointed out, must come to an end somewhere. The problem is that, for anyone not already wedded to the naturalistic objectivist position, explanation seems to have been brought to a halt far too early. Thus, it seems to be perfectly legitimate to ask for an explanation of why a particular property, such as life, richness, and so on, should make something intrinsically valuable. And it seems simply inadequate to say that it just does. The force of the naturalistic fallacy, then, is that naturalistic objectivism brings the scheme of moral explanation and justification to an end at an implausibly early stage of the proceedings. And, thus, naturalistic objectivism seems to be irredeemably arbitrary.

It is important to realise that this point applies quite generally against all forms of naturalistic objectivism, and not just against the environmental versions of Goodpaster and Miller. The same charge of arbitrariness could be made with equal force against any property or set of properties proposed as the ground of intrinsic value. As Callicott points out, without further argument, it seems arbitrary to say, following Kant, that only rational beings are intrinsically valuable because reason is what gives things this value. Similarly, without further argument, it seems arbitrary to say, following Bentham, that only sentient beings are intrinsically valuable because only pleasure is intrinsically good. And, similarly, without further argument, it seems arbitrary to say, following Plato and Leibniz, that only ordered things are intrinsically valuable because order is intrinsically good.[6] At each point, the sceptic is entitled to ask: why is rationality/pleasure/order intrinsically valuable? This question certainly seems legitimate, and if it is, then it requires some sort of explanation of why these features make something intrinsically valuable. The naturalistic objectivist, in other words, stops the explanatory and justificatory process too early in the proceedings.

It might be thought that this problem can be avoided if only the naturalistic objectivist would desist from stopping the explanation and justification of his claims at such a ludicrously early stage. That is, instead of remaining content to assert that, for example, richness is what gives things intrinsic value, the naturalistically inclined objectivist should be prepared to go further, identifying the relevant feature of richness which gives it the ability to make things intrinsically valuable.

However, this strategy would only push the problem back a stage. Suppose the naturalistic objectivist identifies a certain property of richness, call it property P, which, on his view, gives richness the ability to make things intrinsically valuable. Then, he could make claims of the following sort. Richness bestows intrinsic value on the things which possess it because of property P. But, then, the relevant property is not richness itself, but property P possessed by richness. But this means that our objectivist is merely replacing one assertion with another. Instead of claiming that it is richness which makes things intrinsically valuable and leaving the process of explanation and justification at that, he is now claiming that it is property P which makes things intrinsically valuable and leaving the process of explanation and justification at that. But, once again, this seems merely to invite the sceptical retort: why P? What makes P so important?

What the objectivist naturalist seems to require is to identify some property whose connection to intrinsic value is, as we might say, *transparent*. That is, he needs to identify a property whose relation to intrinsic value is so clear, obvious, and necessary, that, by itself, it explains why it makes things intrinsically valuable. He needs to identify a property whose explanatory relevance to intrinsic value is, so to speak, written on its face. This would be a very strange property indeed. No naturalistic theory of value ever succeeded in identifying such a property (if it had, there would be no worries about a naturalistic fallacy). But, even more worrying, it is doubtful that the idea of such a property makes any sense at all. It is doubtful, that is, that to this property there corresponds any coherent concept. And it is only the vacuity, and consequently the mystery, of this property that makes us think it is something we can legitimately search for. In other words, the naturalistic objectivist seems to have put himself in a position that is quite common in philosophy: wanting something he cannot have.

It is worth reiterating that these considerations apply to all versions of this position, and not just the environmental versions of Goodpaster, Miller and others. It is only the familiarity and self-serving nature of more traditional forms of naturalistic objectivism which makes us suppose they have any more inherent plausibility. In isolation from all valuing consciousness, it is just as arbitrary to claim that reason is good and all rational beings are intrinsically valuable as it is to claim that life is good and all living things are intrinsically valuable. It is just as arbitrary to claim that pleasure is good and all sentient beings are intrinsically valuable as it is to claim that richness is good and all sufficiently rich are intrinsically valuable.

This sort of objection is often taken to be more or less fatal to naturalistic versions of objectivism. The objection is, in fact, simply a rehearsal of a point made famous by Hume that there is a logical gap between 'is' and 'ought', such that one cannot derive the latter from the former. However, in the work of the naturalistic objectivist Holmes Rolston III, we find an attempt to bridge this gap.[7] If the attempt is successful, naturalistic objectivism may yet be salvaged; and it is to an examination of this attempt that we now turn.

3.4   Beyond the naturalistic fallacy?

Consider the value possessed by another human being who, in time-honoured philosophical fashion, we can call Jones. What is the source of this value? The value possessed by Jones might, in part, be made up of the fact that she is valued by other people. And, in this regard, we might value her instrumentally – she is a useful member of society, great fun to be with, and so on. Also, we might value her intrinsically; we might recognise that her value exceeds her usefulness for us or society in general, in whatever direction that usefulness might lie. But whether or not this is true, there is still a dimension of her value that we have not yet touched on. Jones values herself, and at least part of her value depends on her valuing herself and derives precisely from this valuing. This aspect of the value of Jones seems, at least *prima facie*, to be both non-instrumental and objective. The value is non-instrumental since it does not derive from Jones' utility either for us or for society as a whole. It would be a strange Jones who valued herself *only* for her usefulness to others. And it seems to be objective in that this aspect of her value does not depend on our beliefs or opinions about, or our attitudes or feelings to, Jones. In this sense, whether or not Jones values herself is a perfectly objective fact about the world: she does or she does not, independently of our beliefs, opinions, attitudes or feelings about the matter. It is this sort of thought that provides the basis for Rolston's attempt to develop a version of objectivism immune to the naturalistic fallacy.

Suppose a plausible case can be made out for the claim that all living creatures value themselves. Then, if some of the value of Jones derives from her valuing herself, then any creature that valued itself would have value, at least in part, in virtue of this fact. And this would seem to be an objective fact about the world. If a creature values itself, then it would do so independently of our beliefs, opinions, attitudes and feelings towards it. So, the question is: can a plausible case be made out for the claim that all living creatures value themselves? And, once we remove certain misconceptions about what is involved in valuing oneself, I think the answer is a qualified yes.

The first misconception consists in the idea that valuing oneself is a conscious phenomenal state of some sort, that is, that it is a state defined by a specific phenomenology. But this is clearly false. Consider the difference between, say, being in pain and believing that Ouagadougou is the capital of the Upper Volta. Being in pain is what is known as an *occurrent* state which means, roughly, that it begins and ends at a specific time and in order to have it you must be conscious of it at all times. But believing that Ouagadougou is the capital of the Upper Volta does not conform to this model at all. It is not an occurrent state but a *dispositional* one, and whether one possesses a dispositional state like this depends not on one's conscious life, but on what one is *disposed* to do in given circumstances: whether one is disposed to answer 'Ouagadougou' when asked what the capital of the Upper Volta is, and so on. The belief, like all others, manifests itself only rarely on the conscious stage. In fact, it manifests itself on my conscious stage only when I am trying to explain the difference between occurrent and dispositional states. But it is nonetheless a belief that I have, that I have had for several years (ever since I read Stephen Stich's *From Folk Psychology to Cognitive Science*, actually), and, presumably will continue to have for some time. Valuing oneself is, similarly, not an occurrent state but a dispositional one. As you read this book, you presumably value yourself, but that you do so is not something that is hovering perpetually in the forefront of your consciousness. Most of the time, it's not hovering in your consciousness at all. It's a dispositional state, something that manifests itself in your dispositions to behaviour.

The second, and perhaps even more common, misconception is that valuing oneself requires that one have a concept of oneself; that in order to value yourself you have to be able to represent yourself to yourself in thought. This misconception derives largely from the mistaken idea that the content of one's thought is always *conceptual* content. We can attribute thoughts and other contentful states to individuals in two very different ways depending, in large part, on the interests we have in making such attributions. Consider the concept of a camera. Possessing this concept, it is plausible to suppose, is a matter of possessing a certain loosely defined set of appropriate beliefs about cameras: beliefs about what cameras are for, how they work, and so on. And in order to conceptually represent or think about a camera, one must possess the concept of a camera. Suppose, now, that a new tribe is encountered, whose members have never seen a camera and who recoil in fear from it. Eventually it is hypothesised that the tribespersons believe that the camera will steal their souls. This hypothesis may, of course, be wrong, but it certainly seems possible that it is true. However, if all content was conceptual content, content mediated by concepts, then this hypothesis would have to be rejected out of court. It could not possibly be true because the natives do not possess the relevant set of beliefs about cameras, therefore do not possess the concept of a camera. If all beliefs had a content that was conceptual, the tribespersons could never believe anything at all about the camera since they do not have the requisite concept. Indeed, neither could they have any desires concerning the camera, since desires are content based states just as much as beliefs.

This, however, seems very implausible. Clearly they do believe something about the camera, otherwise they would not recoil in fear. And clearly they desire that the camera not be pointed in their direction. And what this seems to show is that not all content is conceptual. Sometimes our attribution of a belief or other mental state to an individual is sensitive not to the way they conceptually represent the object of their belief, but to the object itself. This is precisely what is going on when we attribute beliefs about the camera to the tribespersons. Their beliefs, in this case, possess what is known as *de re content*. Their beliefs and desires represent the camera, but they do so non-conceptually. Now just as it is possible to non-conceptually represent external objects such as cameras, so too is it possible to non-conceptually represent oneself. And, if this is correct, then to value oneself, and to desire that one's existence not be abbreviated, does not require that one have a self-concept. One can value oneself non-conceptually.[8]

Now, Rolston does not base his argument on the distinction between occurrent and dispositional states, or on the distinction between conceptual and non-conceptual content, and so the above arguments should not be attributed to him. Nonetheless, they do, I think, remove the biggest obstacles to the claim that creatures other than humans can value themselves. Once we allow that an individual's valuing of itself is a dispositional and not essentially conceptual state, then there seems to be little reason for thinking that this individual must be human. For, if valuing oneself is a dispositional state and one that is not necessarily mediated by a concept of oneself, then the question of whether a given creature values itself is to be determined not by speculation about its likely phenomenological life, nor by reflection on the necessary conditions for possession of a self-concept, but, rather, by reference to its behaviour. Does it avoid doing things that would end its existence? Does it do things that have the purpose of maintaining its existence? If so, then this certainly seems to be good evidence for the claim that it values itself, albeit dispositionally and non-conceptually.

These sorts of considerations, I think, make it plausible to suppose that valuing oneself is something that can be done by all conscious or sentient creatures. But can it be extended beyond the realm of sentience? Can it be extended to living creatures in general? Here, I think, the argument, is less conclusive. Consider, first, what Rolston says about the matter. Rolston points out that all organisms defend their 'own kind as a good kind'.[9] By this he seems to mean that all organisms, even plants, actively defend their lives and strive to propagate their own species. This, I think, is an extremely unfortunate way of developing the argument, raising, as it does, the unnecessary and unwelcome spectre of group selectionism, something we shall have reason to encounter later on also. But, without going into too much detail at present, the claim that all organisms defend their 'own kind as a good kind' seems to overlook that some of the things with which organisms compete are precisely members of their own kind. Now one might argue that this sort of competition between conspecifics might enhance the form or telos of the kind – that the most perfect members of their kind are the ones most likely to be successful in this struggle – but this puts us on the sort of Aristotelian road down which, personally, I do not want to go.

The group selectionist gloss is, in any event, unnecessary. Instead of talking about organisms defending their 'own kind as a good kind', we can simply point out that all organisms actively defend their lives and strive to propagate their *genes*. Animals obviously fight or flee when threatened, and they struggle, sometimes heroically, to propagate their genetic profile. Moreover, plants are not as passive as they appear. They compete with conspecifics for resources, and many secrete allelopathic chemicals from their roots to poison competitors. Does this indicate that every living thing values itself? I can understand the reluctance one might have to grant this. It is one thing to deny that valuing oneself is a state defined by a *specific* phenomenology, but quite another to claim that it can exist in the absence of *any* phenomenology whatsoever, or even the capacity for having a phenomenology. Similarly, it is one thing to claim that valuing oneself is possible in the absence of a *conceptual* representation of oneself, but quite another to claim that valuing oneself is possible in the absence of *any* form of representation.

We have, I think, arrived at one of those interesting points in philosophy where we are simply unclear what we should say. That is, we are in the borderland between what we *would* say and what we *should* say about whether or not an individual values itself. We have reached the limit of what we would say about valuing, therefore we do not know what we should say. Are there any considerations that might indicate what we should say here? I think there are, and they centre around what Callicott has called the *teleological proof* of the existence of intrinsic value.[10] The 'proof' goes like this. The existence of instrumental value entails the existence of intrinsic value. To say that something is instrumentally valuable is to say that it is a means to something else that has value. Now, one means may exist for the sake of another means, but the train of means must, on pain of regress, ultimately terminate in an end which is not a means to anything else. Otherwise, the train of means would be infinite and ungrounded. Thus, the existence of means to an end entails the existence of an end-in-itself. And an end-in-itself is something which has intrinsic value. The argument is essentially Aristotle's, being put forward at the beginning of the *Nichomachean Ethics*.

If we accept this argument, then we have good reasons for saying that all living creatures value themselves. For we are quite willing to accept, I think, that various adapted features of living organisms, features which enable them to survive and propagate, are ones which are of *instrumental* value to the organism, and this is true whether or not the organism is sentient. The deep roots of a desert plant, for example, are instrumentally valuable as means to obtaining ground water. So, there does not seem to be anything semantically perverse in assigning instrumentally valuable functions to adapted features of living organisms, whether they are sentient or not. But, if we accept the above argument, instrumental value entails the existence of intrinsic value. If the deep roots of the desert plant are genuinely instrumentally valuable, then this entails that something must have intrinsic value. It does not, of course, directly entail that the plant itself must have this intrinsic value. But, we are entitled to ask, I think, *if not the plant then what*? Presumably, in this day and age, no one should take seriously the idea that the desert plant is there for the benefit of desert dwelling human beings. So, the instrumental value of the plant's roots is not, without committing oneself to a form of religious implausibility, relative to the intrinsic value of human beings. So, if we allow that various adapted features of the plant can legitimately be said to be instrumentally valuable to the plant, then this strongly suggests, though falls short of logically entailing, that the plant itself has intrinsic value.

To summarise, the argument, as developed so far, looks like this. A striving, self-valuing, organism, whether sentient or not, is an end-in-itself. It, therefore, has a good of its own, and its various adapted features are instrumentally valuable to it in helping it pursue this good. More-over, this good of its own, possessed by an organism, is independent of the beliefs, attitudes, opinions or feelings that *other* valuers bear towards it. And since the possession by an organism of a good of its own does not depend on that organism being conscious, intrinsic value in nature extends beyond the domain of consciousness. If all sentience were obliterated at a stroke, nature would continue to have intrinsic value as long as there exist striving self-valuing individuals. The limits of intrinsic value in nature are the limits of life.

This is, in many respects, an ingenious argument. In broad strokes, we can explain what has gone on here as follows. The charge of arbitrariness that seems fatal to traditional forms of objectivism has been circum-vented by allowing that the value of an individual derives from its being valued. However, what makes the value of an individual objective and intrinsic is the fact that it can derive not from that individual's being valued by others, but from it being valued by *itself*. The value that derives from this self-valuing is, thus, independent of the beliefs, opi-nions, attitudes, feelings and valuing activities in general of *other* indi-viduals. The value is, in this sense, intrinsic and objective. Now, one may question whether we are any longer dealing with a form of objectivism since, on the present account, the valuing of an individual does derive from its being valued, and not from anything else. And this is one instance, I think, where the neat division between objectivist and sub-jectivist accounts of intrinsic value, a division which seemed so neat, decisive and plausible when introduced in Chapter 2, has a tendency to break down. We shall encounter many more instances of this tendency before we are finished. For present purposes, however, we can evaluate the specific proposal of Rolston without worrying about whether it is an objectivist or subjectivist proposal. A rose by any other name, and so on.

What this strategy provides us with, then, is essentially a *biocentric* account of intrinsic value. Nature contains intrinsic value because it contains living organisms that value themselves. Is there any way of extending intrinsic value beyond living things? In at least some of his writings, Rolston seems to think so. He claims that on this view, both species and ecosystems possess intrinsic value. However, his reasons for this are, I think, largely unpersuasive. Consider, first, species. Each organism, according to Rolston, represents its species. Its species, the type of which it is a token, is its essence, its form or telos, and each organism strives to be good-of-its-kind. Because of this, all organisms defend their 'own kind as a good kind'. Hence, since organisms have intrinsic value and since each organism defends its own kind as a good kind, the kind or species itself also has intrinsic value.[11] My stomach, I'm afraid, is simply not strong enough for me to swallow the Aristotelian metaphysical cosmology that is necessary for being persuaded by this argument. The idea that organisms defend their 'own kind as a good kind' is, moreover, incompatible with everything we know about evolutionary theory, which tells us that organisms certainly 'defend their own genes as good genes' but do so by way of often violent, often painful, and sometimes fatal struggle with members of their own kind. And if Rolston were to argue that such struggle only makes its survivors better exemplars of their kind, hence that the struggle indirectly has the result of defending the kind as a good kind, then his argument seems to tacitly rely on the pre-Darwinian assumption of the *immutability of species*. Standard neo-Darwinian theory tells us that the struggle for survival, a struggle which involves conspecifics as well as members of other species, can drive a lineage of organisms from membership of one species into membership of another. Species, in other words, are not immutable. The boundaries between one species and another are not rigid or well-defined, and species are in a constant process of transformation. And this seems to leave little room for seeing evolutionary struggle as a matter of organisms defending their 'own kind as a good kind'.

Rolston's argument for the intrinsic value of ecosystems is equally unconvincing, though for different reasons. The argument seems to go something like this. Ecosystems are the matrix within which each kind or species is formed, where each kind comes to be what it is. But, if each organism defends its own kind as a good kind, then the ecosystem into which that kind, with its value, fits must also be good. If the kind is valuable, then the ecosystem which produces that kind must also have value. Indeed, if the kind is valuable then all the natural processes which led up to the production of that kind – elemental natural processes leading all the way back to the big bang – must also have value.[12] Perhaps, but what kind of value? There seems to be nothing in this argument which suggests anything more than instrumental value for ecosystems. If A has intrinsic value, and B has value only because of its role in producing A, then the value of B seems to be instrumental value. This is virtually a point of definition. Therefore, Rolston seems to have done nothing to support the *intrinsic* value of either species or ecosystems.

In his more recent writings, Rolston seems to acknowledge this.[13] And so we seem driven to the inevitable conclusion that an objectivist account of environmental value can, at most, underwrite the intrinsic value of life. An objectivist account of environmental value, that is, is a *biocentric* account. And if we are to even get this much out of objectivism, much tidying up work needs to be done, perhaps along the lines indicated earlier, concerning what is involved, and, more particularly, what is *not* involved, in an organism's valuing itself. But assuming this work can be successfully accomplished, we have a conclusion of not inconsiderable importance: *life is intrinsically valuable*. And we have arrived at this conclusion in a way that, arguably, is not subject to the charge of arbitrariness predicated on the naturalistic fallacy.

While this is a conclusion of considerable importance, it will not satisfy many, perhaps most, environmentalists. The reason is that many of the most pressing environmental problems are concerned with things that are not living; species and ecosystems being the most obvious examples. We worry about endangered species, but species, as opposed to members of species, are not living things. And we worry about the disappearance of certain ecosystems, but ecosystems, as opposed to some of the things in them, are not living things. So, the question is: can we do any better than biocentrism? Is it possible to develop an account of environmental value that allows for the intrinsic value of species and ecosystems?

If we can, then the way forward might be suggested by our development of the objectivist model. In order to avoid the charge of naturalistic arbitrariness, the strategy adopted by Rolston seems to involve allowing that value does derive from valuing. That is, that the value of a living organism derives from the organism valuing itself. And, it is tempting to suppose, the particularly intimate relation, if we may speak this way, between an organism and itself means that the organism values itself in a particular way: it values itself intrinsically. It cannot value itself instrumentally since to do so it would have to value itself as a means to some further end, and it is difficult to see what this further end could possibly be. And so, one might be tempted to suppose that since the value of the organism derives from the fact that it values itself, and since it possesses intrinsic value, then this is because it values itself intrinsically. The intrinsic value of the living organism derives from the fact that it values itself intrinsically. But if this is the case, then it might turn out that organisms (such as ourselves) might value other things – perhaps species or ecosystems – intrinsically, and this might provide a ground for their possessing intrinsic value. If intrinsic value derives from intrinsic valuing, and *only* from this, then whatever is intrinsically valued might have intrinsic value.

Thus, the arbitrariness inherent in objectivist accounts of environmental value has pushed us in the direction of a subjectivist account of such value. Indeed, it can plausibly be argued, I think, that we have already been pushed over the border of subjectivism. For we have arrived at a view which acknowledges that value can only derive from valuing. And this seems tantamount to subjectivism about value. The arbitrariness of traditional naturalistic forms of objectivism, that is, clearly points the way to, and the need for, a subjectivist account of value. And it is to such an account that we now turn.

4
Subjectivist Theories of Intrinsic Value

In Chapter 2, it was argued, provisionally, that in order to fulfil the function required of it, namely underwriting an environment based, as opposed to a human or sentient based, ethic, the concept of intrinsic value needed to incorporate two features. Firstly, intrinsic value needs to be *non-instrumental* in the sense that it does not derive from, and depend for its existence on, the goals, purposes, and projects of human beings (indeed, of sentient creatures in general). At least some of the value of the natural environment must be independent of the role it can play in furthering human purposes, or the purposes of sentient creatures in general. Without this condition, an environment based ethic would collapse into a human or sentient based ethic, an ethic of environmental management. Secondly, it was argued that, at least at first glance, intrinsic value must be *objective*. At least some of the value of the natural environment must be independent of attitudes of approval, or disapproval, indeed, independent of mental properties in general. Without this condition, the concept of intrinsic value would not be able to satisfy one of the central motivations for its introduction: the protection of the environment in the face of human (or sentient) indifference, or even hostility, towards it. In accordance with these two desiderata, the previous chapter examined the prospects for an objectivist theory of environmental value. However, the central problem with objectivism – the problem of arbitrariness – requires for its resolution that we travel in the direction of, and almost certainly over the border into, subjectivism. Thus, the most plausible version of objectivism turns out almost certainly to not be a version of objectivism at all. And even if it is, this version takes us no further than *biocentrism*: the value of nature is *life*. Life is intrinsically valuable, and this value derives from the valuing by a living creature of itself. On this view, however, anything that is not alive cannot have intrinsic value. And this excludes many core environmental structures, notably species and ecosystems. And, to many environmentalists, this is unacceptable. Indeed, in the eyes of many, it is structures such as species and ecosystems, as opposed to individual organisms, that are the primary bearers of value. So, it is time to examine if we can do better than a biocentric account.

The biocentric account of environmental value avoided the problem of arbitrariness by accepting that the value of the environment must derive from its being valued by someone or something. And this valuing consisted in a relation that a living organism bears to *itself*. The intrinsic value of the living organism, that is, derives from its valuing itself non-instrumentally. It is possible, however, for a living, valuing, organism to value things other than itself; in particular, it might value species and ecosystems. Of course, if we regard the value of these things as deriving purely from their being valued by something else, then we have not probably but definitely abandoned an objectivist model of environmental value in favour of a *subjectivist* one. However, some environmental philosophers have argued that this is quite acceptable. The need for the second condition on the concept of intrinsic value, the objectivity condition, has been challenged. The basis of this challenge is that all the work required of intrinsic value can be accomplished simply by accommodating the first condition, that of non-instrumentality. This idea has received probably its most influential and complete theoretical articulations in the work of J. Baird Callicott and Robert Elliot. This chapter provides a critical evaluation of their arguments.

4.1　Elliot's indexical model of intrinsic value

Robert Elliot has developed what he calls an *indexical* theory of intrinsic value.[1] This theory is subjectivist in that it makes the value of the environment dependent on certain mental properties of human beings. However, Elliot also claims that his account of value is non-instrumentalist, in that it does not make the value of the environment consist solely in its role in furthering human purposes, and realist, in the sense that it makes at least some statements about environmental value true. And this, Elliot argues, is sufficient to underwrite the core claims of environmental ethics. This section outlines Elliot's theory.

Objects, states of affairs, events, processes, actions etc. can, according to Elliot, all possess intrinsic value. If they do so, however, they will do so in virtue of possessing some other properties, which Elliot calls *value adding properties.* Value adding properties are ones which add to the intrinsic value of items. It does not follow from the fact that something possesses value adding properties that it automatically possesses intrinsic value, for it may also possess value subtracting properties. Nevertheless, if an object has value adding properties, then it is more valuable than it would be if it did not possess such properties.

Value adding properties need not be, and usually are not, essential properties of a thing. Nor need they be simple unanalysable properties. Most importantly, value adding properties need not be intrinsic properties of things to which they add value. The value that such properties add can be intrinsic without the properties themselves being intrinsic. Many value adding properties, therefore, might be *relational* properties of things; the property of being rare and that of being naturally evolved provide good examples here.

Properties are value adding, according to Elliot, if 'they exemplify the property of standing in the approval relation to an attitudinal framework. In fact, the property of being value adding is identical with the property of standing in the approval relation to an attitudinal framework.'[2] This claim will take some unpacking, but its central idea is quite clear: roughly speaking, something has intrinsic value if it is approved of by a valuer in virtue of certain properties that it has.

Suppose, for example, that George says 'X has intrinsic value'. Then this claim will be true only if George approves of X. Whether anyone else approves of X is irrelevant. Similarly, when Georgina says 'X has intrinsic value', her claim will be true only if Georgina approves of X. Whether George or anyone else approves of X is irrelevant. That is, judgements that something has intrinsic value are indexical in character: 'they are indexed or relativised to the valuer making the judgement, the time at which the judgement is made and the world in which it is made.'[3] The claim 'X has intrinsic value', that is, is really a shorthand form of a much more specific claim: 'X is approved of by a particular valuer at a particular time and in a particular possible world.' Indexing to times is introduced because the attitudes of a particular valuer might alter over time. What has intrinsic value for George at one time might not have at a different time. Indexing to possible worlds is introduced to accommodate the possibility that George's attitudes might have been other than they are.

Intrinsic value is, therefore, relative to particular valuers and particular times (and particular worlds). There is no such thing as intrinsic value as such, but only intrinsic value for George, intrinsic value for Georgina, and so on for each valuer. However, intrinsic value might not, in practice, be so fragmented since many valuers will share the same value judgements concerning objects, and even if all the value judgements of one valuer do not precisely match those of another, there may still be large domains of overlap. This is why Elliot uses the notion of an attitudinal framework in the characterisation of his theory. An attitudinal framework, here, is simply the set of attitudes or psychological states that a valuer bears towards the world around her. Distinct valuers can share the same attitudinal framework, if not precisely then at least to a greater or lesser extent. Thus, when George says 'X has intrinsic value', intrinsic value is, strictly speaking, indexed to George, but it is also, more loosely, indexed to his attitudinal framework and, therefore, indexed to all those who share George's attitudinal framework (or at least that part of his attitudinal framework that is relevant to his valuing of X).

Therefore, applying Elliot's model to the environment, the idea is that whether or not the environment has intrinsic value is relative to particular valuers at particular times in particular worlds. The claim that the environment has intrinsic value is a claim with no clear meaning. To the extent that it has meaning it must be understood as an elliptical form of a claim such as 'The environment is intrinsically valued by a particular person at a particular time.' And that, on Elliot's view, is all that the intrinsic value of the environment can be.

4.2   Problems with Elliot's theory

Elliot has provided a subjectivist model of intrinsic value, a model based quite centrally on the psychological states of valuers. Intrinsic value is to be built up out of intrinsic *valuing*: the intrinsic value of the environment is to be constructed out of a certain type of valuing of the environment by human subjects. Will this work? I think, in fact, Elliot's account faces formidable difficulties.

The central difficulty concerns whether Elliot's conception of intrinsic value is sufficiently robust to serve the purposes of an environment centred ethic. There are two issues to consider here. The first is whether Elliot's theory can allow for the claim that statements about the value of the environment can be true or false. Suppose, for example, that deep ecologist Georgina claims that the environment has intrinsic value and should, therefore, be protected. Anti-environmentalist George, however, disagrees, and believes we should treat the environment in a way that most effectively promotes human interests. How are we to adjudicate between George and Georgina? Well, intuitively, any prospect of defending Georgina's environment centred position requires, at the very least, that her statement about the environment having intrinsic value be potentially true. If there is at least the prospect of her statement being true, then we can set about trying to prove that it is, or at least finding reasons which suggest that it is true. If we know at the outset, however, that there is no prospect of her statement being true, then there is simply no point in trying to prove or support her statement, and, it seems, we must abandon the attempt to rationally persuade our anti-environmentalist opponents of the validity of our position. Victory in the dispute under these circumstances will simply be a matter of who shouts loudest.

The second issue concerns the possibility of rational discussion and evaluation of environmentalist and anti-environmentalist claims. The truth or otherwise of such claims is, at least intuitively, not sufficient for the rational discussion of such claims for there can be true claims for which no rational evidence can be amassed, and false claims against which no rational evidence can be brought to bear. Before the advent of space travel, for example, people could make perfectly meaningful claims about the nature of the far side of the moon. Some of these claims were true, some were not. But there was, at that time, no rational evidence to adjudicate between many of these claims. It seems that any attempt to develop an adequate environment centred ethic needs to make room for the claim that it is possible to rationally defend the claims of such an ethic. There is simply no point in attempting to develop an environment centred ethic if the claims that make up this ethic are not ones that can be rationally defended. This is so because, as we saw in Chapter 2, the rationale for developing an environment centred ethic was precisely in order to have a means of rationally defending the environment in the face of human hostility or indifference.

Therefore, I think it is fair to say that if Elliot's account is unable to underwrite both the truth of intrinsic value claims about the environment and the rational defensibility of such claims, then Elliot has failed to construct an adequate account of intrinsic value. I shall argue that Elliot clearly fails to satisfy the requirement of rational defensibility. Adopting Elliot's theory means that there is no genuine scope for the rational defence of claims of environmental value. Whether or not his theory satisfies the condition of truth is less clear cut. Arguably it does, but only in a way that emphasises and reinforces his failure to account for rational defensibility. Therefore, Elliot's model of intrinsic value is inadequate for the purposes of an environment centred ethic.

Consider, first, the question of the truth of intrinsic value claims about the environment. Elliot is quick to characterise his theory as a *realist* account of intrinsic value. And he claims that it is realist because it entails that judgements about intrinsic value take truth-values. Moreover, he also claims that his account is not relativist, where this is understood (in, I think it is fair to say, a very unorthodox way) as the view that two contradictory assertions can have the same truth-value. Elliot claims that when George says 'X has intrinsic value' and Georgina says 'X does not have intrinsic value' they are not making contradictory claims. The reason this is so is that each claim has a truth-condition that is appropriate to an indexical sentence. Thus, for example, when George says 'I am here' and Georgina says 'I am here', their utterances can be both true and non-contradictory. Thus it is with intrinsic value claims.

So, Elliot's position seems to be this. When George says at a particular time t 'The environment has intrinsic value' the truth-conditions of what he says are 'George intrinsically values the environment E at time t'. And when Georgina says, again at time t, 'The environment does not have intrinsic value', the truth-conditions of what she says are 'Georgina does not intrinsically value the environment E at time t'. The claims purport to be about the environment, but what makes them true are facts about George and Georgina. Nevertheless, on Elliot's view, claims that the environment has intrinsic value are ones which can be true or false; it is just what makes them true or false are facts about valuing individuals not facts about the environment itself.

The problems with this account begin to emerge when we consider Elliot's response to a possible objection. At some time in the distant past, say one hundred million years ago, there were, let us suppose, no valuers. Nevertheless, one might want to claim that the environment then had intrinsic value in virtue of, for example, its biodiversity. But, it seems that Elliot's indexical theory of intrinsic value rules out this attribution. When there were no valuers, then, it might seem, there was no value, at least on Elliot's theory. Elliot, however, is quick to deny that this follows. The reason is that even though there were no existing valuers one hundred million years ago, nonetheless today there are valuers, and the valuers of today value the environment of one hundred million years ago. So, the environment of one hundred million years ago possesses intrinsic value in virtue of being valued by the valuers of today.[4]

This shows, however, that judgements of intrinsic value can, for Elliot, involve time in two different senses. First, they are indexed to times in that such judgements are made by an individual at a particular time. When, George, for example, at a particular time t, says that the environment has intrinsic value, then George's judgement is thereby indexed to t. However, George can also make judgements not only about the environment in general, but also about the environment at particular times. Thus, he can value the environment not only of today, but also of one hundred million years ago. So, when George says now that the environment of one hundred million years ago has intrinsic value, his judgement of intrinsic value involves time in two senses. It is *indexed* to the present in that he makes the judgement now. And it is *directed towards* one hundred million years ago, since it is a judgement about the environment of one hundred million years ago.

Given that this is so, consider the following scenario. George, at a particular time t_1, values the environment of that time intrinsically and, at this time, makes an utterance to this effect. Thus, we can say:

George intrinsically values (E at t_1) at t_1

Thus we can say that the environment at t_1 has intrinsic value for George at t_1. However, George at a later time t_2, changes his mind about the value of the environment. All his previous thoughts about the intrinsic value of the environment he now comes to regard as mistaken. And this revision of his attitudes applies also to his earlier valuings of the environment. Thus, whereas before he thought that the environment had intrinsic value at t_1, he now thinks that this earlier belief of his was mistaken. Thus, we can say:

George does not intrinsically value (E at t_1) at t_2

Thus, we can say that the environment at t_1 does not have value for George at t_2. But then the question arises, does the environment have value for George at t_1 or not? And we seem to be committed to the claim that the environment both does and does not have value for George at t_1. We seem, on Elliot's indexical theory, to be committed both to the claim that the environment does and that it does not possess intrinsic value for George at t_1.

This objection is not fatal. Elliot can avoid it by pointing out that just as judgements of environmental value are indexed to particular times, so too judgements that the environment has a particular value at a particular time are indexed to particular times. This, however, leads directly to the second problem with his account. If judgements that the environment has intrinsic value at a particular time can be indexed to times, then it is possible, in principle, for one and the same valuer, in the course of a lifetime, to make literally millions of conflicting judgements about the value of the environment even at a particular time. At t_1 he believes the environment has intrinsic value at t_1, at t_2, however, he changes his mind and believes that it does not have intrinsic value at t_1. At t_3 he changes his mind again and believes that it does have intrinsic value at t_1, and so on. In fact, our hypothetical valuer can vacillate throughout the course of his life without, on Elliot's view, him *ever believing anything contradictory* about the environment. That is, Elliot's attempt to account for the truth of claims of environmental value leads him to adopt a standard of contradiction according to which the above sort of vacillation involves no contradiction.

Not only does this seem implausible, more importantly it highlights the problems Elliot is going to have with the rational defensibility of claims of environmental value. If our hypothetical valuer can continually vacillate between the claim that the environment at a particular time has intrinsic value and the claim that the environment, at the same time, does not have intrinsic value *without ever contradicting himself*, then how is rational evaluation of his claims possible? If a basic logical device such as contradiction is not applicable *even here*, then it is difficult to see upon what basis rational evaluation of his claims might take place. More generally, the problem is that, on Elliot's view, whether or not the environment has intrinsic value depends on the attitudes valuers bear towards it. And, beyond certain limits, the having of an attitude is not something that is subject to rational evaluation. You either have the attitude or you do not. Bedrock is reached before canons of rationality can get adequate purchase.

Elliot disputes this claim. He argues that his indexical theory does allow for the possibility of rational discussion and engagement. This is so, he argues, for several reasons. Firstly, a valuer can be mistaken about his or her self-indexed judgements.[5] So, when George claims that X has value (for George), Georgina may be able to show him that his judgement is false. X does not have value for George, he just mistakenly thought that it did. Thus, for example, George might intrinsically value X because he believes it to have the property of being naturally evolved. His belief that being naturally evolved is a value adding property might be correct, but his belief that X has this property might be false. And the latter belief is something over which George and Georgina might rationally disagree. Alternatively, George might not intrinsically value X because he has failed to notice certain value adding properties, and Georgina can bring these to his attention. There can also be disagreement about the precise nature of value adding properties. George and Georgina can, for example, agree that sentience is a value adding property, but disagree about the nature of sentience. In all these ways, then, George and Georgina can rationally engage.

Secondly, Elliot argues that rational evaluation of value judgements is possible within the framework of the indexical theory because there is a consistency constraint on such judgements.[6] Pointing out inconsistent attributions of value, then, can rationally compel a change in judgement.

Third, rational engagement is made possible by the fact that attitudinal frameworks are open-ended and incomplete. Because of this, there will be properties that each valuer will not have considered as candidates for value adding properties.[7] These are not properties which are value adding but which have gone unnoticed by a particular valuer; rather they are properties which the valuer has never considered as candidates for value adding properties. Offering such properties for the consideration of those with whom one is disputing, Elliot claims, counts as rational engagement.

Elliot is correct, I think, in his claim that his indexical account of intrinsic value can allow for a *degree* of rational evaluation of value judgements. But the crucial question is whether it allows for a *sufficient* amount of rational evaluation for the purposes of an environment centred ethic. And, I think, the answer is that it clearly does not. The basic problem with Elliot's attempt to make room for rational discussion and evaluation within the framework of his indexical theory of value is that it does not adequately cover what we might refer to as *deep* disagreements over what has value. And it is these sorts of disagreements that characterise disputes over the value of the environment. Let us consider each of Elliot's arguments in turn.

Firstly, Elliot claims that we might be mistaken over whether or not an object possesses a value adding property, and rational argument can rectify this sort of mistake. This is true, but it does not address the sort of situation that is the most problematic aspect of environmental ethics. This is the situation where all disputants agree over what properties a given object of discussion possesses, but disagree over whether these properties are value adding. Elliot claims that there can be disagreements over the precise nature of value adding properties, and rational discussion can be useful in clarifying and resolving this sort of problem. This is again true. However, this sort of disagreement does not characterise the central disputes of environmental ethics. The fundamental disputes in environmental ethics, for example, do not arise over the precise characterisation of what it means for something to be naturally evolved, or for it to be rare. Nor do they arise from disputes over whether an object actually possesses such properties. The disputes arise because there is fundamental disagreement about whether these sorts of properties are value adding or not.

Secondly, Elliot also claims that rational discussion and evaluation is possible within the framework of the indexical theory because there is a consistency constraint on attributions of intrinsic value. Pointing out inconsistent value judgements, then, is a means of rationally compelling a change of judgement. However, there are two problems with this argument. Firstly, as we have seen, Elliot's theory has the consequence that the statement 'The environment has intrinsic value at time t' is not inconsistent with the statement 'The environment does not have intrinsic value at t'. And, if this is true, it is difficult to say the least, to what criteria of consistency Elliot is appealing. Secondly, conditions of consistency will only be sufficient to decide questions of environmental value if it can be assumed that either those who assert, or those who deny, the intrinsic value of the environment are being inconsistent. But it would be a brave environmentalist indeed, I think, who based his case for the intrinsic value of the environment on the claim that those who deny such value are being logically inconsistent. Wrong they *may* be, though that has yet to be shown. But it is unlikely, to say the least, that they are being logically inconsistent.

Elliot's third argument is that it is possible for there to be candidates for value adding properties which, as yet, have remained unconsidered, and it is possible to put these to one's disputant as candidates for value adding properties. This, for Elliot, is rational engagement. However, once again, if this is rational engagement, then it is so only in a way that is too weak for the purposes of an environment centred ethic. In defending the intrinsic value of the environment one can put forward candidates for value adding properties until one is blue in the face. This will not help in the rational development of an environment centred ethic unless there is some means of rationally compelling acceptance, by one's disputant, of at least one of the proffered candidates.

The root problem with each of Elliot's arguments, and, indeed, with his indexical theory of value in general, is that while it may allow for *some* rational discussion, evaluation and argument of core environmentalist value claims, it does not allow for *sufficient* rational discussion, evaluation and argument for these core environmentalist value claims to be defended against someone who does not share them. Therefore, I think we should conclude that Elliot's model of intrinsic value is not sufficiently robust for the purposes of an environment centred ethic.

4.3 Callicott's truncated model of intrinsic value

In the work of J. Baird Callicott, it is possible to discern two distinct approaches to the concept of intrinsic value. One of these fits fairly comfortably within what, in Chapter 1, I referred to as a traditional approach to environmental value. The other is clearly radical in character.[8] Callicott's radical approach to environmental value will be discussed in Chapter 5. This chapter is concerned with his version of the traditional approach.

Callicott's traditional approach to intrinsic value is based on an innovative coupling of three distinct strands: the account of moral judgement associated with the empiricist philosopher David Hume, the evolutionary account of moral sentiments developed by Charles Darwin, and the environmental or *land* ethic of Aldo Leopold. Consider, first, Hume's account of moral judgement.

Hume's account of morality is sometimes referred to as a form of *sentimentalism*.[9] By this is meant a certain model of the origin and nature of moral judgements. According to Hume, a moral judgement such as, for example, that a certain action is good or that it is evil, is not founded upon reason but upon sentiment. Good and evil are not objective qualities of things in the world, rather they derive from us. In particular, good and evil are feelings of approval or disapproval, approbation or disapprobation, sympathy or repugnance, that actions and events elicit in us. Hume illustrates his account with a now famous example. Suppose we come upon the scene of a murder. Upon inspecting the body, we might be able to discover how long the person has been dead, the cause of their death, what the murder weapon was, we might even be able to identify the murderer's motive. These are all what Hume would call *matters of fact*. But that the murder is morally wrong is something we can never discover by looking at the world. The wrongness of murder is not a matter of fact, it is not a fact objectively contained in the murder. Rather, it is something we bring to the murder; something that we *project* onto the murder. When we come across the murder scene, we feel, perhaps, horror, dread, aversion, panic, alarm, and so on. And we project these on the murder. The feelings cause us to judge that the murder is wrong, and the projection of these feelings upon the murder itself causes us to think of this wrongness as a fact objectively contained in the murder as such. Moral judgements are the result of feelings, or *sentiments*, that we project onto external events. And it is the projection of these sentiments which causes us to think of the moral judgement as representing a feature of these external events.

Callicott accepts the Humean account of the origin of moral judgements. He accepts, therefore, that all value originates in a valuing subject. Value, in this sense, has a subjective origin. However, he points out that to make a claim about the origin of value is not, at least not directly, to say anything about which objects possess value. In particular, to say that the origin of value lies in various subjective feelings or sentiments of human beings (or sentient creatures in general) does *not* entail that the only things which possess value are those feelings or sentiments. To suppose that it does would be to confuse the *origin* of a judgement of value – the reason why a judgement is made – with the *content* of a judgement of value – with what that judgement is about. And even if it is true that the value of a thing derives from certain sentiments that we project upon it, it is also true that we can value things in two very different ways. Some things we value instrumentally; we value them for some other good which we think they will help us obtain. Money and medicine are obvious examples. However, other things we do not seem to value in this way. Parents do not value their children, for example, in the way in which they value money or medicine. Parents are supposed to value their children, and in non-pathological cases do so value them, in a way that is independent of any selfish goal of the parent. Parents, that is, do not value their children simply as things to amuse them, as things to love, as things to provide for them when they are old and infirm. Although a child may indeed have instrumental value for a parent, its primary value is independent of its instrumental character.

Generalising this point, Callicott argues we can value nature in the same sort of way that a parent values the child. The value we attribute nature is, of course, attributed by us, but it is not instrumental value; it is not value that depends on any contribution that nature might make to our goals and purposes. We can allow that the value of all things, nature included, has a subjective origin in the feelings or sentiments of humans. Value, in this sense, is *anthropogenic*. But this does not entail that all value is *anthropocentric*, at least not if this is understood as the claim that the only things of value are human beings or human sentiments. Nature has a value whose origin lies in the sentiments of human beings. Nevertheless, this value is non-instrumental and is genuinely possessed by nature, or, at least, as genuinely possessed by nature as any other relational property.

The second strand of Callicott's account of environmental value consists in his incorporation of certain elements of Darwinian thought, in particular, Darwin's account of the evolution of the social instincts.

Callicott's adoption of Hume's sentimentalist account of moral judgements might be thought to lead directly to a form of chaotic moral relativism. If value depends on feelings of sentiment, and if such feelings vary wildly from one person to another, then what has value will fluctuate accordingly. Hume was quick to deny this possibility on the grounds that the moral sentiments – sympathy, empathy, fellow-feeling, affection, and the like – are natural and universal features of human beings, possessed by all *normal* (i.e. non-sociopathic) individuals. And, if this is so, moral judgements will not vary wildly from one person to another. However, Hume's claim here is certainly contestable, and what is required to make it plausible is some account of *why* the moral sentiments should be naturally and universally distributed among humans. This is where Callicott's appropriation of Darwin comes in.

According to Darwin, the origin of morality lies in the parental and filial affections common, probably, to all mammals.[10] That is, it lies in what in the present day is referred to as *kin altruism*. Bonds of affection between parents and offspring have a respectable evolutionary explanation in that such bonds can promote the survival of the genes passed on from one to the other. Such bonds permit the formation of small social groups based on kinship (i.e. sharing of genetic profile). However, suppose a random mutation, occurring in certain individuals but not others, resulted in the extension of the natural parental and filial affections to less closely related individuals. This would permit the enlargement of the family group. And should the newly extended community be better at surviving than smaller groups, for example by being more adept at defending itself, or more adept at provisioning for itself, then the fitness of its individual members would also be increased. Hence the mutation would be passed on at a differentially greater rate. And in this way, these more diffuse parental and filial affections, which Darwin, echoing Hume, calls the *social sentiments*, would spread throughout a population. In this sort of way, Darwin argued, the moral sentiments could have co-evolved with the evolution of proto-human societies. The moral sentiments were necessary to the continuing survival of such societies, and the societies, in turn, conferred differential inclusive fitness on their members.

Therefore, the reason the moral sentiments do not vary wildly from one person to another is that they are evolutionary products. They are not superficial, or whimsical features of us, and we cannot change them as we would a jacket. They are engrained into us as part of our natural history. Those who do not possess them, we label sociopaths, or psychopaths, and we regard them as very much abnormal cases.

The third strand of Callicott's account of environmental value consists in the incorporation of certain elements of Aldo Leopold's environmental or, as he calls it, *land* ethic. The Humean strand of Callicott's account explains how moral sentiments provide the basis of our moral judgements. The Darwinian strand explains how we come to possess these sentiments and how this possession should be uniform, or near uniform, across all normal human beings. The function of the third, Leopoldian, strand, is to explain how these uniformly distributed sentiments can be environmentally directed; that is, how they can be applied to, and provide the basis for, our judgements about the value of the natural environment. The central idea, here, is of a cognitively driven *redirection* of the basic moral sentiments built into us through our evolutionary history. For while our possession of moral sentiments is, ultimately, an evolutionary, hence innate, matter, what these sentiments are directed towards is flexible and shaped by learning. The Darwinian account of the moral sentiments explains why we should have moral sentiments directed towards members of our own community. But ecology, according to Leopold, 'simply enlarges the boundaries of the community to include soils, waters, plants, and animals, or collectively: the land.'[11] Ecology allows us to appreciate that we are embedded in, and constituted by, the natural environment just as much as the social. The natural world is, from an ecologically informed perspective, a living whole. The myriad species, which, in our ecological naivety, we might regard as extrinsically, contingently, and haphazardly related are, from the ecological perspective, seen to be intimately conjoined, specifically adapted to each other, to the soil, water, the climate, to the *land*. And we human beings exist within this natural community. We are, in short, citizens not only of human communities but also of a *biotic community*.[12]

Combining the three strands, Humean, Darwinian, and Leopoldian, gives us what Callicott calls a *truncated* account of environmental value. Our judgements about what is right or wrong, what has value and what does not, and what has intrinsic value and what has only instrumental value derive, as Hume pointed out, ultimately from our moral sentiments. These moral sentiments, as Darwin argued, are distributed uniformly, or near uniformly, across all normal human beings. And ecological awareness, understanding of our position within a genuinely biotic community, can make us see the natural environment as an appropriate object for the direction of our moral sentiments. Therefore, according to Callicott, the natural environment has a value whose origin lies in the feelings or sentiments of human beings. However, this value, while anthropogenic, is not anthropocentric: the value of the environment does not consist solely in its instrumental role for us because, while we might, in part, value it instrumentally, we do not value it *only* in this way. At least to some extent, we can value the environment as a good in itself. And we can value it in this way because evolution has built into us certain moral sentiments which cause us to value things as intrinsically valuable, and these sentiments, via the sort of cognitive redirection occasioned by familiarisation with the principles of ecology, can be directed towards the natural environment.

4.4 Problems with Callicott's view

It is now quite common to find Callicott's position objected to on the grounds that it fails to capture the *normative force* of moral judgements.[13] The problem with basing a theory of moral judgement on the possession – uniform or otherwise – of certain moral sentiments is that such possession is a descriptive rather than normative fact about the world. That is, the Humean-Darwinian account of the moral sentiments developed and endorsed by Callicott can at most give us an account of what sentiments human individuals *as a matter of fact have*. But it cannot supply an account of what sentiments human individuals *should* have. But moral judgement, it might be argued, necessarily goes beyond what is; it is concerned with what should be. Moral judgement, that is, is essentially *prescriptive* rather than *descriptive*, but Callicott's account is merely *descriptive*.

I think this is, to an extent, a legitimate concern, but will not push the point here. This is partly because Callicott has attempted to deal with it elsewhere, although I think it is fair to say that the jury is still out on the success of this attempt.[14] But, perhaps more importantly, I think there is something fundamentally unfair with levelling this objection specifically at Callicott. The same objection applies equally to any naturalistic moral theory. Any naturalistic meta-ethical theory will always, ultimately, be a descriptive, rather than a prescriptive, account of the world. Now, the adequacy of naturalistic accounts of morality in general, and their success in capturing the normative dimension of moral judgement in particular, are both legitimate concerns. But it seems, to say the least, a little unfair to expect Callicott to sort this one out. The relative merits of naturalistic and non-naturalistic accounts of morality seems to be a discussion that is best situated elsewhere.

More worrying perhaps is that Callicott's use of the notion of cognitive redirection, the Leopoldian strand of his argument, seems to steer him perilously close to the view that the value of nature is, after all, only instrumental value. In developing the idea of a cognitive redirection of naturally evolved moral sentiments, Callicott introduces the distinction between what he calls *ultimate* and *proximate* human values.[15] Ultimate human values are 'values about which there is little or no disagreement or debate', certain 'basic values [that] are universal characteristics of human nature'. Proximate values, on the other hand, are 'values about which there is very often a great deal of disagreement and debate.'[16] According to Callicott, which are and are not the correct proximate values depends on being right about the facts. The reason we value things proximately is that we believe they will promote what we value ultimately. Thus, 'our debates about proximate values are debates about means, not ends.' Since what we value proximately depends on our knowledge of certain facts – facts pertaining to the relation between those things we value proximately and those things we value ultimately – this allows for the possibility that our ultimate values, whose existence is constituted simply by our possessing certain moral sentiments, can be redirected towards proximately valued things by way of relevant factual considerations. For example, health is, presumably, an ultimate value of normal human beings. But, at various times, people have had a variety of ideas about what will proximately contribute to this ultimate value: mineral baths, leeches, and, in a more contemporary vein, colonic irrigation, have all been thought by some people to proximately contribute towards the ultimate value of health. So, what proximate values we hold will be essentially a matter of what we believe are the facts, specifically, the facts about what will promote our ultimate values. In a similar, but undoubtedly more complex way, knowledge of environmental facts – facts which are often complex and presuppose significant understanding of ecological theory – can cognitively redirect our shared ultimate values in an environmental direction.

What is most striking here is Callicott's assimilation of proximate values to means, and ultimate values to ends. For this seems to entail that the environment is a simply a proximate means to promoting our ultimate, shared, human ends. And this seems to be equivalent to claiming that the environment has only instrumental value. Or, to put the same point another way, what exactly is it that knowledge of ecological theory provides us with? One fairly obvious answer is that it provides us with knowledge of just how much we depend on our environment, how we have been shaped by it, both physiognomically and psychologically, and how much our well being, hence our shared ultimate values, depend on it. But if this is what ecology tells us, then the cognitive redirection occasioned by familiarisation with ecological theory seems to amount to nothing more than the idea that the environment is important because it is essential to our well being. And if this is correct, then Callicott's account of environmental value is an instrumental account.

Callicott, of course, clearly seems to have more than this in mind. And what seems to be involved is something like Leopold's idea that familiarisation with the principles of ecology engenders in us 'a sense of kinship with other creatures; a wish to live and let live; [and] a sense of wonder over the magnitude and duration of the biotic enterprise.'[17] Now it is clear that ecology *might* promote such feelings in us, but it is less than clear that it *would*. And so, I think it is fair to say that Callicott's case is incomplete here. What he seems to require is not only an account of the uniform distribution of moral sentiments, and hence the ultimate values they collectively constitute, but also an account of why we should uniformly, or near uniformly, react to a given set of ecological facts in a certain way; that is, in the way Callicott's account requires us to react if it is not to collapse into a form of instrumentalism about environmental value. And it is, to say the least, unclear what form such an account would take.

The third objection that can be raised against Callicott is, I think, logically the most serious. The objection is that Callicott's theory is internally flawed. In particular, Callicott has got it wrong about the evolution of the moral sentiments, and this makes it impossible to apply his account to the environment via a Leopoldian inspired cognitive redirection of such sentiments. More precisely, there are two questions which need to be considered. Firstly, does Callicott provide a plausible account of the possession of the moral sentiments generally and of the role these play in moral judgement? Secondly, is it plausible to suppose that Callicott's account of the moral sentiments can provide the basis for an environment based ethic? The first question asks whether Callicott has got it right about the moral sentiments in general. The second asks whether Callicott can legitimately apply his account of moral sentiments, even if correct, to the case of the environment. I shall argue that Callicott is *almost* right about the evolution of moral sentiments generally. However, in a small but crucial respect his account is mistaken. And putting his account right means that it cannot then plausibly be extended to the case of the environment. A correct account of the evolution of the moral sentiments makes problematic any attempt to use them to supply a basis for an environment centred ethic.

Consider, first, Callicott's account of the moral sentiments in general. The first point to note is that Callicott's appropriation of Darwin's account of the evolution of the moral or social sentiments has serious problems, even when viewed from within a broadly Darwinian framework. The major problem, here, is that the explanation is what is known as *group selectionist*. The concept of group selectionism is not unambiguous. In one sense, for example, it is understood as the claim that natural selection operates at the level of groups, so that at least one of the types of entity that are selected for by natural selection are groups of individuals. Callicott's position is not, at least not necessarily, group selectionist in this sense. However, there is another sense of group selectionism, and in this latter sense Callicott's view does qualify as group selectionist.[18] A position is group selectionist in this second sense when it explains the existence of a group in terms of the claim that the fitness of an individual can be increased by adopting behaviour which contributes to the survival of the group. The problem with this behaviour is that it does not, for the individual, constitute what is known as an *evolutionary stable strategy*.[19] Therefore, such behaviour would actually be selected against. This will all take some explaining, which is perhaps best done by way of a simple example.

Consider a population of a hypothetical species whose members exhibit two kinds of fighting strategy: *hawk* strategies and *dove* strategies.[20] The designations 'hawk' and 'dove' here, of course, refer not to types of bird but to behavioural characteristics possessed by individuals of any species. Let us suppose that any individual of our hypothetical population is classified either as a hawk or a dove. Hawks always fight as belligerently as they can, giving up only when seriously injured. Doves merely threaten their opponents by way of conventional gestures and actions. Thus, if a hawk fights a dove, the dove runs away as quickly as possible. If a hawk fights a hawk they continue until one of them is seriously injured or dead. If a dove meets a dove, nobody gets hurt, they simply go on posturing at each other until one of them tires and so backs down.

Now, as a purely arbitrary convention, we can allot the contestants points, which we can think of as directly convertible into the currency of gene survival. For example, we might allot 50 points for a win, 0 for losing, -100 for being seriously injured, and -10 for wasting time. The specific numerical values are not important. An individual who scores high points is an individual who leaves a greater number of genes in the gene pool than an individual with a lower score.

Suppose, now, that we have a population consisting exclusively of doves. Whenever they fight, nobody gets hurt. The contests consist simply of prolonged ritual tournaments which end only when one rival gives up. In these cases, the winner scores 50 points for gaining the resources in the dispute, but pays a penalty of −10 for the time he has wasted in the dispute. So, he scores 40 in all. The loser is also penalised −10 for wasting time. On average, any one individual can be expected to win half his contests and lose half. Therefore, his average pay-off per contest is the average of +40 and −10, which is +15. Therefore, each individual dove in a population of doves seems to be doing quite nicely.

However, suppose now that a mutant hawk arises in the population. Since he is the only hawk, every fight he has is against a dove, and doves always run away. Therefore, the hawk scores +50 every fight. Thus, he enjoys an enormous genetic advantage over the doves whose average pay-off is +15. And, hawk genes will, therefore, rapidly spread through the population. However, once this happens, each hawk can no longer count on every rival he meets being a dove. To take an extreme example, if the hawk gene spread so successfully that the entire population came to consist of hawks, all fights would now be between hawks. But things would then be very different. When a hawk meets a hawk, one of them is seriously injured, scoring −100, while the winner scores +50. Each hawk in a population of hawks can expect to win half his fights and lose half his fights. His average expected pay-off is, therefore, the average of +50 and −100, which is −25. Now consider a single dove in a population of hawks. To be sure he loses all his fights, but on the other hand he never gets hurt. His average pay-off is 0 in a population of hawks, whereas the average pay-off for a hawk in a population of hawks is −25. Dove genes will therefore tend to spread through the population.

The way the story has been told so far suggests that there will be a constant oscillation between hawk and dove genes in the population. However, it need not be like this. There is a stable ratio of hawks to doves. For the particular arbitrary points system we are using the stable ratio would be 5 doves to every 7 hawks. When this stable ratio is reached, the average pay-off for hawks is the same as the average pay-off for doves. If there are oscillations about this stable point, they need not be large ones.

It may be thought that I have prejudiced matters with the particular numerical assignment of gains and losses I have chosen. But this is not so. Essentially the same result will be achieved for any realistic assignment of values. The problems with group selection should, therefore, hopefully now become clear. The average pay-off to an individual in a stable population with a ratio of 5:7 doves to hawks is 6.25. This is true whether the individual is a hawk or a dove. Now 6.25 is much less than the average pay-off for a dove in a population of doves, which is 15. If only everybody could agree to be doves, every single individual would benefit. Group selectionism would, therefore, predict a tendency to evolve towards an all dove group, since a group which contained a 7/12 population of hawks would be less successful. But the trouble with such an all dove conspiracy, even though it is to everyone's advantage, is that it is fatally open to abuse. In an all dove group, a single hawk does so well that nothing could stop the evolution of hawks. The group is, therefore, bound to broken by, so to speak, treachery from within.

The problems with Callicott's appropriation of Darwin should, now, also be evident. A group made up exclusively of individuals who have evolved to possess feelings of affection, sympathy, benevolence, etc. towards their fellow members – a group of altruists, let us say – is not an evolutionarily stable group. It is fatally susceptible to invasion by a mutant egoist who does not possess these feelings. The mutant egoist gene would, therefore, spread through the population. Indeed, we should, in principle, be able to predict the stable ratio of altruists to egoists. Evolutionary theory, therefore, cannot explain the emergence of moral sentiments distributed universally and naturally throughout a group, with individuals not possessing such sentiments stigmatised as *abnormal*. On the contrary, evolutionary theory predicts that any such group will possess, as biologically normal members, both individuals who do and individuals who do not possess such sentiments. Thus, Callicott's appropriation of Darwin is too simplistic.

In fact, the picture painted so far is itself too simplistic. It is very implausible to suppose that evolution has worked by equipping certain individuals with the full array of moral sentiments and by equipping others with none. It would be an unlikely eventuality, that is, to find society divided into pure egoists and pure altruists. It would be far more realistic to suppose that society is made up of individuals who, to a greater or lesser degree, incline more to the egoistic end of the moral spectrum, and individuals who, to a greater or lesser extent, incline to the altruistic end. This too is compatible with the requirements of evolutionary stability outlined above. In any event, evolutionary theory, when properly understood, predicts that a community should be made up of individuals who differ, often quite markedly, with respect to the moral sentiments they possess. The idea that evolution has equipped everyone, except *abnormal* sociopaths and psychopaths, with the full range of moral sentiments rests on a misunderstanding of evolutionary theory.

This point can be supported by the following thought. Suppose Callicott was correct about evolution equipping all normal members of society with the whole gamut of moral sentiments. If this were so, we would have a hard time explaining exactly why there were so many moral problems in the world. Why would people often be wicked, self-ish, thoughtlessly egocentric if everyone, or at least all normal people, had been equipped by evolution with the full range of moral senti-ments? If evolution were capable of ensuring a near universal distribu-tion of moral sentiments throughout every member of society who was not abnormal, then we would be hard pressed to explain the manifest *lack* of sympathy, benevolence, affection, fellow feeling, among large numbers of the population. The answer, as we have seen, is that even if you assume that the moral sentiments are purely evolutionary products, then evolutionary theory does not predict that these sentiments will be universally distributed among normal members of the population. On the contrary, evolutionary theory predicts that there will be a stable ratio of individuals who possess the sentiments to a greater extent to those who possess them to a lesser extent.

Standard evolutionary explanations of altruism do not now presup-pose that evolution has ensured a near universal distribution of moral sentiments throughout all normal members of human society. Rather, they assume that evolution, by itself, will yield a distribution of such sentiments that is partial, and that is not uniform across different indi-viduals. The stable ratio of egoists to altruists is brought about by differ-ent individuals possessing, to different extents, both egoistic and altruistic qualities. Given this is so, how do we explain the near uni-formity of moral action? If we have different moral endowments, differ-ing amounts of egoistic and altruistic proclivities, how do we explain the great degree of uniformity in our moral judgements and actions? Murderers, as Callicott is correct to point out, are very much abnormal cases. If a near universal distribution of moral sentiments faces the problem of explaining the amount of moral wickedness in the world, it seems that anything much less than this will have the converse problem of explaining why people with radically different natural moral endowments almost always toe the line with regard to moral judgement and action.

The answer evolutionary theory provides us is based around the notion of *sanction*. Sanction takes up the slack; fills in the gap that evolutionary theory predicts should exist between our sentiments and our actions. Living as we do in social groups, it is almost always in our own best interests to toe the line as far as our actions, if not our senti-ments, are concerned. It is in our interests, for example, to persuade our conspecifics that we are people who can be trusted, people with whom one can safely do business, and so on. And convincing our conspecifics of this will often involve performing actions that are of no short-term benefit to us. We engage in apparently self-sacrificing behaviour not because we are altruists, but, ultimately, because we are rational egoists. However, to this fairly traditional line of thought, recent evolutionary accounts add a twist.[21] Engaging in rationally self-interested behaviour is not always, or even usually, an explicit consciously calculating activity. The same sort of behaviour is found in virtually all social mammals, including those for whom it is unrealistic to suppose they have a complex conceptual grasp of the future, hence of future benefits. Rather, the explanation of this behaviour seems to be that individuals who are predisposed to this sort of self-sacrificing behaviour, whether or not they realise it is in their long-term interests, can, in the context of a social group at least, be selected for over individuals that are not. And, crucially, it does not matter *why* the individual is disposed to engage in this ostensibly self-sacrificing but covertly self-enhancing behaviour. Feelings of sympathy, benevolence, affection, fellow-feeling, and so on would be perfectly acceptable facilitators. This is indeed altruism. But altruistic activities have a firm foundation on a bedrock of egoism.

This recent evolutionary account is based on the notion of sanction in the following sense. The idea is that individuals should be in a position to punish, or in other ways take sanctions against, any conspecific who repeatedly acts through naked self-interest against the interests of others. It is essential to the account that there be differential treatment of individuals who engage in self-sacrificing behaviour and those who do not. Self-sacrificing behaviour, whether intentional or not, must be rewarded; nakedly self-enhancing behaviour must be discouraged. The account, then, works only if there is a mechanism for discriminating between self-sacrificing and self-enhancing behaviour, and for reinforcing the former and discouraging the latter. In the context of a social group, it is clear that there would be such a mechanism, one that lies in the discriminating responses of conspecifics.

However, and this is crucial, in the case of the environment, there is no such sanctioning mechanism. Of course, if we act in a shortsighted and ill-advised manner towards the environment, then we are likely to suffer the consequences. But these consequences will be felt equally both by those who exploit the environment and those who cherish it. If one person messes up the environment, everyone suffers. And, therefore, the environment cannot act as a sanctioning mechanism.

So, this is the problem Callicott faces. His account of the evolution of the moral sentiments is group selectionist and, therefore, cannot be correct. This is not because Callicott has misunderstood Darwin. On the contrary, he has understood him all too well. Darwin himself did not grasp the distinction between individual and group selectionism, therefore he did not appreciate the problems with his account of the evolution of the moral sentiments. More recent evolutionary theory, however, predicts a gap between our moral sentiments, which undoubtedly exist but are non-uniformly distributed throughout the population, and our judgements and actions which clearly are almost uniform. And the slack, here, is taken up by the mechanism of sanction. However, in the case of the environment, no relevant sanctioning mechanism is available. Therefore, it seems that Callicott's account of the evolution of the moral sentiments is incorrect in a small but crucial way. And when we correct this, we see that his account cannot underwrite our intrinsic valuing of the environment.

5
Radical Approaches to the Value of Nature

The theories of intrinsic value examined in Chapters 3 and 4 are examples of what I have referred to as *traditional* approaches. What makes an approach to environmental value traditional is a certain picture or conception of how this value must be. The conception takes the form of a disjunction, an *either/or*. Either the value of nature must be subjective in that it depends for its existence and nature on the mental activities of human, or at least conscious, valuers, or it must be objective in that its existence and nature are independent of the mental activities of conscious valuers. And what motivates this orienting conception of how value must be is a further, and therefore more basic, distinction between *subject* and *object*.

Consider what is involved in consciously thinking about something, for example, having the thought that the sky is blue. On the one hand, there is the object that you are thinking about, on the other the act of thinking itself. The former is an object of consciousness, an object of thought. This is not to say that it is *only* an object of consciousness, that it only exists when someone is conscious of it. Objects of consciousness, or thought, can, of course, typically exist independently of their being thought about. To say that something is an object of consciousness or thought, here, is simply to make the mundane claim that it is an object of which we are conscious, or about which we are thinking. On the other hand, there is the act of thinking, as opposed to the object thought about. And this latter act is generally regarded as part of the subject of consciousness. Therefore, reality, it seems intuitively natural to say, can be divided up into two sorts of things. On the one hand, there are things which belong to, or are part of, the objects of consciousness, on the other there are things that belong to, or are part of, the subjects of consciousness. This simple picture is complicated somewhat by the existence of introspection which, on certain construals, involves consciousness becoming an object for itself, but this is not something we need worry about. However, once we separate the world in this way into subjects and objects of consciousness, the picture of value as either subjective or objective is made inevitable. Once we separate the world into subjects and objects, that is, we must then ask ourselves where value is to be found: is it to be found among the subjects of consciousness or among its objects? Subjectivist theories of the value of nature are based on the idea that this value is, ultimately, located among the subjects of consciousness. Environmental value, that is, is, in some way, a product of the valuing activities of conscious subjects. Objectivist theories of natural value, on the other hand, are based on the idea that the value of nature is, ultimately, to be found among the objects of consciousness. Environmental value is not something constructed by the mental activities of valuing subjects, but, rather, is located out there in the world, something whose existence and nature is independent of these valuing activities. It is something, therefore, which can be discovered or detected by way of human mental activities – it can be an object of consciousness – but not something that is constituted or constructed by such activities.

The picture of environmental value as either something subjective or as something objective, then, depends on the distinction between subjects and objects of consciousness. Traditional approaches to the value of nature, therefore, presuppose this distinction. Radical approaches, on the other hand, deny it. Traditional approaches presuppose the division of the world into subjects and objects of consciousness; radical approaches deny the legitimacy of this division.

5.1 Environmental value and quantum theory

In Chapter 3, we examined the subjectivist theory of environmental value developed by Callicott. In some of his writings, however, Callicott abandons a straightforward subjectivist account of the value of nature in favour of an approach that is distinctly more radical. The inspiration and justification for this shift is twofold, provided, on the one hand, by developments in quantum physics: the branch of physics that deals with the nature and behaviour of sub-atomic particles, and, on the other, by ecological theory, the science that deals with the terrestrial environment as a whole (as opposed to some part of it).[1] Callicott argues for a claim (also expressed by Warwick Fox), that ecological theory and certain interpretations of quantum theory provide 'structurally similar' or analogous representations of environmental systems and micro-physical nature respectively.[2] That is, the picture of reality contained in certain interpretations of the quantum theory is similar, or analogous, to the picture of reality implicated in ecological theory. Examination of quantum theory, then, can perhaps provide us with a model for thinking about environmental value also. This section and the next focus on Callicott's treatment of quantum theory. The remaining sections deal with his appropriation of ecological theory.

The crucial feature of quantum theory, the feature which makes it significant for the purposes of environmental theory, is that, on Callicott's view it negates, or renders illegitimate, the subject/object distinction (and the fact/value distinction predicated upon it). It does this by undermining the illusion of the passive, non-participating, observer. At the meso- and macrolevels of phenomena, that is, at the levels of billiard balls, planets and stars to which the investigations of early modern physics were largely confined, the illusion of a wholly passive, non-participating observer can be maintained. However, when we turn to extremely small-scale phenomena – protons, neutrons, electrons, and below – this illusion collapses. The reason is that, in order to observe an object, it is necessary to exchange energy with it. In order to see the black cat that is hiding in your coal shed at midnight, you must shine a torch on it. Now, this energy exchange is so small that, in everyday contexts, it can be ignored. The beam of light that you project on the cat is of such low magnitude compared to the cat, that it will not appreciably change the nature of the cat (since the mass of the cat is equivalent to a vast amount of energy). However, when the object of observation is very small – of sub-atomic stature or below – then the act of observation will contain enough energy that it will disrupt the behaviour, and even the nature, of the object in appreciable ways.

For example, according to Heisenberg's famous uncertainty principle, one can never have precise knowledge of both the velocity and the position of an electron. Knowledge of one of these quantities automatically rules out knowledge of the other. That is, by measuring the position of the electron, one makes knowledge of its velocity indeterminate. And, by measuring its velocity, one makes knowledge of its position indeterminate. The energy possessed by certain sub-atomic entities and the time during which they exist are indeterminate in the same sort of way. Knowledge of one quantity automatically and necessarily rules out knowledge of the other. Moreover, it is generally regarded as impermissible in normal scientific contexts to posit the existence of entities or quantities that cannot in principle be observed. Therefore, according to what is known as the Copenhagen interpretation of the quantum theory – which Callicott thinks is the most famous and still most influential interpretation of this theory – the uncertainty of velocity/position and energy/time has ontological as well as epistemological consequences. Thus, it is claimed that when we measure, for example, the position of an electron we render indeterminate not only our knowledge of its velocity but also its actual position. Not only can we never know its exact velocity, it does not actually have an exact velocity. Similarly, when we measure its velocity, the electron does not have an exact position. And the reason it does not is precisely because our observation of its velocity somehow makes it have no position.

In this way, the act of observation actually changes the nature of the object observed. And this, if correct, undermines the illusion of the passive observer and, thereby, breaks down the division between subject and object. On this interpretation of quantum theory, the nature of an object observed (an object of consciousness) is dependent on the act of observation (the subject of consciousness). Thus, at least in quantum contexts, it is impossible to legitimately separate subjects and objects of consciousness. The nature of quantum objects depends essentially on acts of observing subjects.

On the basis of these sorts of considerations, Callicott develops a radical model of environmental value. Traditional approaches to value distinguish between what are known as *primary* and *secondary* qualities. According to the philosopher John Locke, one of the early advocates of the distinction, an object's primary qualities are its mass, its location, velocity, and so on. Its secondary qualities include such features as colour, flavour, odour, and so on. Primary qualities are traditionally thought of as objective in that whether or not an object has them does not depend on the mental activities of conscious subjects. Secondary qualities are traditionally thought of as subjective in that they are not independent of the mental activities of subjects in this way. Quantum physics, on Callicott's view, breaks down the distinction between primary and secondary qualities: 'Location and velocity are potential properties of an electron variously actualised in various experimental situations as colour and flavour are potential properties of an apple awaiting the eye and tongue (and all the neural apparatus that goes with eyes and tongues) of a conscious being for their realization.'[3]

Callicott recommends that we understand values in the same sort of way. Value in general, and the value of nature in particular, is a kind of secondary quality. Thus, the value of nature is not intrinsic, nor is it objective and independent of consciousness. However, to concede this, Callicott argues, is to concede nothing of consequence, since quantum physics shows that *no* properties are strictly intrinsic. No properties, that are, are objective in the sense that their existence and nature are independent of consciousness. Instead of a misplaced quest for objectivity, we should understand values as *virtual*.[4] It is not absolutely clear what Callicott has in mind here, but the general idea seems to be something like this. Contained in nature is a store of values that exist potentially but not actually. They come to exist actually only when a valuing consciousness interacts with them and, in doing so, actualises them. Some of these virtual or potential values are instrumental in character, that is, valued for their utility as material-economic resources or psycho-spiritual resources. And some of these values are inherent in character: the things that have them are valued for their own sake. In this way, Callicott grafts the distinction between instrumental and inherent valuing, a distinction originally developed within the framework of his subjectivist model of value, onto a new quantum inspired model of value that is supposedly neither subjectivist nor objectivist. And what is essential to this new radical model of value is the idea that, 'inherent value is a virtual value in nature actualised upon interaction with consciousness.'[5] This radical model, then, 'puts values on an ontological par with other properties including culturally revered quantitative properties... Physics and ethics are, in other words, equally descriptive of nature.'[6]

Callicott's subjectivist model of value regarded value as projected onto the world by valuing subjects. That model tacitly presupposes that the world itself exists objectively; it is required in order for there to be something upon which value is projected. Callicott's new, quantum inspired approach, however, rejects this second assumption. All properties are now virtual properties. All properties, both value properties and quantitative properties, exist potentially in nature, but they exist actually only when encountered and, thereby, actualised, by consciousness.

5.2   Problems with Callicott's theory

The problems with Callicott's account can be broken down into four, not entirely distinct, categories. The first objection is that it is difficult to see how to extend conclusions derived from reflections on quantum levels of reality to the sort of meso-levels that are of concern to environmental thought. The second is that he misunderstands and/or misappropriates quantum theory. The third is that his account still retains a tacit commitment to the subject/object distinction. And the fourth is that Callicott breaks down the distinction between subject and object in a way that is inimical to the purposes of environmental thought. These categories are not entirely separable and categories three and four in particular are, I think, closely related.

The first category of objection to Callicott's model, one upon which I shall not dwell for any length of time, concerns the possibility of extrapolating conclusions applicable to quantum levels of reality to the sort of meso-levels that, presumably, are the ones of interest to the environmentalist. If the position and velocity of a lepton cannot both be accurately determined, it is nonetheless true that both quantities can be determined for medium sized objects like cars, or missiles. Or, if it is claimed that they cannot, it is uncertain what standard of accuracy is being used. If both the energy and time of occurrence of a photon cannot both be accurately and simultaneously determined, it is nonetheless true that the energy and time of an emission from an electric light bulb can both be simultaneously determined. Whatever consequences quantum theory has for understanding of the ordinary everyday, medium-sized world, these consequences are, to say the least, not obvious. Therefore, it is unclear how we can legitimately extrapolate conclusions applicable to quantum levels of reality to larger scale structures such as the natural environment.

Consider, now, the second category of objection. Again, I do not propose to dwell on this type of objection at any great length. Briefly, however, two points can be made against Callicott here. Firstly, it is perhaps worth pointing out that the Copenhagen interpretation of quantum theory is not the most widely accepted and influential interpretation. Not any more. Indeed, the whole idea that there is such a thing as *the* Copenhagen interpretation is itself problematic, since the writings of Bohr, the principal architect of 'the' interpretation, are systematically ambiguous. Some of his claims suggest a 'no fact of the matter' view of what is going on at the quantum level. Others, however, involve the idea of a collapse of a quantum wave superposition when a measurement is made. These interpretations are often taken to be equivalent, but they are far from that. And what is taken to be 'the' Copenhagen interpretation of quantum theory is often a mélange of these two distinct and incompatible views.

Secondly, and more importantly, there is nothing in quantum theory, under whatever interpretation you adopt, which entails the sort of blanket assertion that all properties are, in Callicott's terminology, *virtual*. Even 'the' Copenhagen interpretation of the quantum theory does not, as is commonly thought, entail that all properties of the world are dependent on human consciousness for their existence and nature. Indeed, this can be seen from following the logic of Callicott's argument to its conclusion. Consider the Heisenberg uncertainty principle, of which Callicott makes significant use in arriving at his eventual conclusion. According to Callicott, the uncertainty principle shows that things such as leptons do not have both a precise position and velocity, and which one of these quantities they actually (as opposed to virtually) have is determined by our acts of observation. Observing its velocity actually gives the lepton no precise position, and vice versa. Therefore, what properties leptons actually have depends on our acts of conscious observation. Does this give us reason for thinking that all properties are virtual in that they depend for their existence on acts of consciousness? On the contrary, it gives us a reason for thinking the opposite. For now it is true – objectively true – that leptons possess the property of having velocity and position related in the way specified by the uncertainty principle. That the velocity and position of a lepton should be related in the way specified by this principle is a perfectly objective fact about the world; it does not, in any way, depend on an act of consciousness for its truth. Indeed, I think it would not be inaccurate to say that the objectivity quantum theory subtracts from entities and quantities, it pays back in terms of the objectivity of *relations* between entities and quantities. Quantum theory may give us a relational view of reality, but it does not give us an anti-objective one.

More precisely, within the framework of quantum theory, the state of a system has to be expressed as a *wave function* or *state vector*. And here, the relevant properties of objects cannot be expressed in simple values (as when we say the velocity of a car is 60 mph), but must be expressed as a combination of such values. This is what is meant by saying that a quantum state is a *superposition* of simpler states. Perhaps the simplest example is the property of *spin*. In quantum theory, the spin of a particle has two basic values, labelled 'up' and 'down'. However, in general, the spin of a particle can only be expressed as a *combination* of up and down. Spin is, thus, not like a more familiar quantity such as 'clockwise' or 'anticlockwise' for here we can combine clockwise and anticlockwise moments to yield a simple value of so many degrees clockwise or so many degrees anticlockwise (90 degrees clockwise plus 45 degrees anticlockwise yields 45 degrees clockwise, etc.). Spin is not like this: 'up' and 'down' components are not separable, at least not in this sort of way. Instead, each particle's spin has to be regarded as a superposition of a spin-up state and a spin-down state. The same goes for position and velocity, except that each of these has an infinite number of basic values. The position of a quantum particle must, therefore, be expressed as an infinite-dimensional vector, with a different magnitude for each of its locations. And this vector is best regarded as a *wave*, with different amplitudes at different locations in space. Similarly, the velocity of a quantum particle can be regarded as a wave with different amplitudes at different basic values of momentum. Now, because these states are just vectors, they can be decomposed into components in many ways. While it is often useful to see a two-dimensional spin vector as a sum of up and down components, it can be decomposed in many other ways depending on the basis upon which the vector space is chosen. Now, the relevant point for our purposes is that while, according to quantum theory, the state of a subatomic item is best expressed by a wave function rather than discrete quantities, this wave function is perfectly determinate and perfectly objective. The behaviour of the wave function – how it evolves under (almost) all circumstances – is given by the Schrodinger wave equation, and this a determinate, and indeed (contrary to popular misconception) *deterministic*, description of a perfectly objective reality. It is not lack of objectivity that quantum theory promotes so much as a relocation of objectivity from its traditional loci. Quantum theory tells us that on a basic level reality is wavelike. Objectivity has not, therefore, been lost, but simply relocated: it now resides in wave functions rather than discrete entities. Quantum theory admittedly gives us a very different view of reality than classical mechanics: it gives us a very strange view of reality, but it does not give us an essentially anti-objectivist one.

Let us turn now to the third category of objection to Callicott's appropriation of quantum theory. The problem here seems to be that Callicott's account retains a tacit commitment to the subject/object dichotomy. Consciousness, in Callicott's quantum-inspired model, is given a new role, but it is still interpreted in essentially subject/object terms. To see this, recall Callicott's account of value. Nature is a store of values that exist not actually, but only potentially. The values embodied in nature are *virtual* values: they exist fully or properly only when actualised by an encounter with human consciousness. But, if we focus on the subject/object distinction, it is difficult to see any significant difference between this and the traditional view. Callicott's model, that is, seems to retain, in all essential respects, the subject/object dichotomy. According to the traditional view, at least in caricature, the world exists 'out there', and consciousness exists 'in here'. Consciousness, however, can go 'out there' and encounter various things: objects, events, properties, and the like. Now the question arises as to whether values are among the things that consciousness encounters on its journey. Objectivists say they are, subjectivists deny this. Subjectivists say that values really belong to the nature of consciousness, and that consciousness sometimes gets confused on this score and thinks that what is really part of it is located 'out there'. This is, of course, caricature; but, for our present purposes, it will do. The question is, to what extent this caricature changes in Callicott's new model. Well, in fact, all that seems to change is what consciousness does on its wanderings. Instead of merely encountering things that exist 'out there', it now plays a role in determining the existence and nature of these things. But that is all. But this seems to be not so much a change in the essential nature of consciousness as simply a change in what it does. Instead of merely encountering objects, it now, in part, constitutes them. But the picture of consciousness as essentially an *interiority* remains. Instead of venturing out and encountering objects, it now ventures out, encounters virtual properties, and constitutes them as actual properties. But this is simply to envisage a slight change in the terms of the subject/object relation, it is not to break down the distinction. What is required for this latter task is a reconceptualisation of the *structure* of consciousness itself, and not simply a reconstrual of the *function* of consciousness. Callicott, in other words, has, at most, dealt the subject/object distinction a glancing blow. The distinction survives.

The final category of objection to Callicott's view is closely related to the third. The objection is that Callicott's attempted demolition of the subject/object distinction is inimical to environmental thought. The basis of this objection, is that Callicott attempts to break down the distinction from the wrong direction. His appropriation of quantum physics is aimed at showing the extent to which the world is dependent, for its existence and nature, on the activity of consciousness. Consciousness is what brings the various features of the world – from quantitative properties to values – from virtual status to actuality. And this movement of thought makes the world dependent on consciousness. Callicott's argument, therefore, continues a long tradition of idealist and neo-idealist thought about the environment. I argued in the opening chapter that there are, in fact, two ways of breaking down the subject/object distinction. There is the familiar neo-idealist, or *humanist*, strategy of *pulling the world into the mind*: making the world dependent for some part of its existence and nature on the structuring activity of the mind. On the other hand, there is what I called, perhaps somewhat tendentiously, the *environmentalist* strategy of *pulling the mind into the world*. The environmentalist strategy proceeds by trying to show that while the mind may indeed structure the world, as humanism claims, this is of only secondary importance because the mind is itself *worldly*: the mind itself possesses environmental constituents, it is made up, in part, of environmental structures. Exactly how this is so will not really become evident before the development of what I refer to as an environmentalist model of the mind, and this is one of the tasks of the second half of this book. For now, it is perhaps sufficient to point out that the humanist attempt to break down the subject/object distinction, of which Callicott's theory is a specific version, is inimical to environmental thought to the extent that it makes the existence and nature of the world dependent on consciousness. The environment is, thus, ontologically dependent on consciousness. One might, therefore, naturally expect the value of the environment to be dependent on the value of consciousness. Hargrove has detailed the ways in which philosophy has failed the environment in failing to come adequately to grips with its existence, and hence its value. Humanism, neo-Kantianism in a broad sense, is surely philosophy's greatest failure in this regard. The remainder of this book, in effect, is to indicate, in admittedly broad strokes, how philosophy might attempt to rectify this failure.

Callicott's employment of quantum theory does not end with the Copenhagen interpretation. Callicott also points out that certain, more controversial, interpretations of quantum theory also lend support to environmental thought, but do so in a somewhat different way. Here, however, his case parallels an argument he derives from ecological theory, rather than quantum physics. Much the same sorts of considerations, and much the same sorts of problems, arise from Callicott's treatment of ecological theory and his treatment of the more controversial interpretations of quantum theory. These do not, therefore, require separate discussion. And, in the following sections, I shall focus on Callicott's argument from ecological theory.

5.3 Environmental value and ecological theory

The other inspiration for Callicott's attempt to dismantle the subject/object distinction is provided by ecological theory. According to Callicott, the classical (i.e. pre-quantum, pre-ecological) conception of nature can be presented, at least in broad caricature, as follows: The natural environment consists of a collection of individual objects or bodies, each of which is a collection of atoms. A living body is an essentially mechanical device whose generation, development, decay and demise can be explained in atomistic, mechanical terms. Some of these mechanical devices are inhabited by a 'ghost in the machine'. All of these objects, however, are essentially monadic in that they bear no essential connection to each other (apart, perhaps, from various relations of classification: phyla, class, genera, species, and the like).

Ecological science, for Callicott, undermines this classical picture of nature. The conception of nature implicated in ecological theory is not atomistic and mechanical but *organic*; a picture whose theoretical superstructure is founded not upon atoms and the void but upon *energy*. According to the conception of nature supplied by ecological theory, individual organisms are best viewed as *dissipative structures*. Any living organism persists only if there is a continual flow of energy into it. Living individuals exist only as local disturbances or structurings of this universal flow of energy. We might imagine this energy flow as a stream of flowing water.[7] A living individual is like a vortex in this stream. The vortex does not exist as an entity in its own right, but only as a process: only as a continually changing group of water molecules. And should the flow cease, the vortex necessarily disappears. What we regard as individual things are, in a similar vein, just temporary structurings of the flow in a universal energy field. There is nothing in this picture, of course, that denies the reality of atoms and molecules in the classical science, rather it claims that these things are themselves just temporary structurings in the shifting of the universal energy field.

This organic conception of nature is, Callicott argues, also *holistic*. If living organisms are structurings in a complex field of flow patterns, then it is impossible for them to exist, indeed impossible for them to be conceived of, apart from this energy matrix of which they are modes. Indeed, following Naess,[8] Callicott even suggests that the ecological picture, in effect, revives the doctrine of internal relations associated with turn of the century idealism: the doctrine, very roughly, that the essence of any supposedly individual thing is determined by its relations to other things, and cannot be conceived of independently of these relations. In Callicott's hand, the revival of the doctrine by ecological theory takes an essentially evolutionary form. Callicott's point here is so important as to be worth quoting:

> From the perspective of modern biology, species adapt to a niche in an ecosystem. Their actual relationships to other organisms (to predators, to prey, to parasites, and disease organisms) and to physical and chemical conditions (to temperature, radiation, salinity, wind, soil, and water pH) literally sculpt their outward forms, their metabolic, physiological and reproductive processes, and even their

psychological and mental capacities. A specimen is, in effect, a summation of its species' historical, adaptive relationship to the environment...a species has the particular characteristics that it has because those characteristics result from its adaptation to a niche in an *ecosystem*.[9]

Now, unlike Callicott, I would not want to regard the idea expressed here as amounting to anything like a revival of the doctrine of internal relations. Nevertheless, there is in this passage, I think, a point of profound significance both in terms of the assault on the subject/object distinction, and in terms of the development of a proper conception of environmental value. Its true significance is, perhaps, not properly appreciated by Callicott, and certainly not properly utilised. Indeed, working out the consequences of this idea is, in effect, the task of the later chapters of this book.

5.4   Ecological theory and the subject/object distinction

According to Callicott, the dismantling of the subject/object distinction by way of quantum theory is of a piece with the dismantling that proceeds by way of ecological theory. 'Ecology and contemporary physics', he writes, 'complement one another conceptually and converge towards the same metaphysical notions ... [they] draw mutually consistent and mutually supporting abstract pictures of nature ...'.[10] If we focus just on the Copenhagen interpretation of quantum theory, however, then I think that this claim is simply false. Using the expressions introduced in the opening chapter, Callicott's employment of quantum theory works by *pulling the world into the mind*, whereas his use of ecological theory works by *pulling the mind into the world*. The dismantling of the subject/object distinction that is (allegedly) brought about by way of quantum theory is simply the latest in a long line of humanist, neo-Kantian, problematisations of that distinction. The attack on the distinction occasioned by ecological theory is something very different; it is, I shall try to show, a properly environmental mode of thought.

Callicott's utilisation of quantum theory, I have argued, suffers from two problems pertinent to present concerns. Firstly, it attempts to break down the subject/object distinction in a way that is inimical to environmental thought. Callicott's appropriation of quantum theory has the goal of showing the extent to which the world is dependent, for its existence and nature, on the activity of consciousness. Consciousness is what brings various features of the world from virtual status to actuality. But a movement of thought of this kind is essentially idealist. Secondly, Callicott's quantum inspired attack does not dismantle the subject/object distinction but, at most, delivers it a glancing blow. Instead of merely encountering things that exist 'out there', consciousness now plays a role in determining the existence and nature of these things. But this seems to be not so much a change in the essential nature of consciousness but simply a change in what it does. Instead of venturing out and encountering items, it now ventures out, encounters virtual items, and constitutes them as actual. But the picture of consciousness as *interiority* remains. What is required to properly break down the subject/object distinction is a reconceptualisation of consciousness itself, and not a relatively minor alteration in the terms of the subject/object relation. On this reconceptualisation, consciousness does not encounter objects, virtual or otherwise, nor does it constitute them. It is already among its objects. More precisely, it is constituted by these objects.

The second of these objections is, I think, logically prior to the first. It is precisely because of his failure to properly deconstruct the conception of consciousness as an interiority which encounters/constitutes the world that Callicott is left with a particular conception of dismantling of the subject/object distinction. And according to this conception, dismantling the distinction is a matter of showing how the world is dependent on consciousness. If we leave the traditional conception of the mind alone, then it cannot be pulled into the world because it is, on that conception, essentially an interiority. And the only way, therefore, of breaking down the subject/object distinction is by pulling the world into the mind. Idealism, then, derives from the traditional Cartesian conception of the mind as an interiority.

This is why Callicott's appropriation of ecological theory is so important. For in it we find, I think, the roots of a dismantling of the traditional conception of consciousness itself. In it also, therefore, we find the roots of a genuinely environmental dismantling of the subject/object distinction. In Callicott's use of ecological theory, of course, there is no talk of consciousness constituting the world; there is no attempt to make the existence and nature of the world dependent on the mind. Rather, human beings, we are told, in common with all living things, are simply structurings in the field of energy. More importantly, as the quoted passage makes clear, in the case of adapted creatures, not only their outward form, but also their internal structure and functioning, has been literally sculpted by their relations to the environment.

Most importantly, however, the same holds true of their mental capacities. The mental capacities of any evolved individual, in common with all its other capacities and features, are a 'summation of its species historical, adaptive relationship to the environment.'

In developing this idea, Callicott cites with approval a passage from the ecologist Paul Shephard. The passage is itself instructive and worth quoting.

> Internal complexity, as the mind of a primate, is an extension of natural complexity, measured by the variety of plants and animals and the variety of nerve cells – organic extensions of each other. The exuberance of kinds [is] the setting in which a good mind could evolve (to deal with a complex world). . . . The idea of natural complexity as a counterpart to human intricacy is essential to an ecology of man.[11]

This passage is instructive because it is ambiguous. There are two ways in which it might be construed; one which accords closely with the traditional view of the mind as interiority, the other suggestive of a much more radical view. The traditional construal would be that the human mind has, because of the complexity of the environment, evolved in such a way that it reflects this complexity. Here the guiding conception is of the mind as a mirror of the world, and the interaction between the two is one of causal transaction: the complexity of the mind is caused by the complexity of the world. The more radical construal is that the evolution of the human mind has proceeded by way of *utilising* the complexity of the world. More precisely, the idea is that, in its evolution, the human mind employed various complex structures that were to be found in its environment. That is, the human brain evolved in such a way as to meet the demands of the environment by using various structures in that environment. The capacities of the brain would than have evolved in conjunction with capacities to employ – to manipulate and exploit – relevant structures in the environment. Mental processes, then, processes subserved by the brain, would involve, partly but essentially, these environmental structures. The primary relation between mind and world, on this more radical construal, would not be one of causal transaction but of *constitution*. The mind would not just causally interact with the world; it would, quite literally, be composed of the world.

If this sort of radical story can be made to work, then the mind would be worldly; it would be environmentally constituted. The processes which constitute the mind would themselves be constituted by environmental objects and structures. And, then, the traditional Cartesian conception of the mind would have been dismantled. I shall argue that this sort of story can, indeed, be made to work. Showing precisely how it works will be the task of the following chapters. Developing an account of environmental value on the basis of this conception will be the task of Chapter 9.

5.5   From ecological theory to environmental value?

The significance and promise of Callicott's appropriation of ecological theory has hopefully been made clear. However, even if we are in possession of an environmentally legitimate way of dismantling the subject/object distinction, there is still no straightforward inference to an adequate conception of environmental value. At this point, it is perhaps worth taking time to consider Callicott's suggested move (or apparently suggested move) from ecological theory to environmental value.

If living organisms are simply temporary structurings or patterns in the flow of energy, then there is no essential distinction between an organism and its environment. A vortex in a stream of water is, in an important sense, constituted by the water around it: it could not exist without the surrounding water and cannot be conceived without this water. Similarly, if 'the very structure of one's psyche and rational faculties are formed through adaptive interaction with the ecological organisation of nature, then one's self, both physically and psychologically, merges in a gradient from its central core outwardly into the environment.'[12] However, if there is no essential distinction between human beings and the environment they inhabit, then this allows a type of extension of egoistic thinking. This is important because, as Callicott and others have plausibly argued, egoism is usually taken to be axiologically primitive.[13] Most systems of ethics begin by taking the intrinsic value of oneself as a given, and then seek to generate moral commitments to others on this basis. The usual argument for this extension involves the idea that there are no morally relevant differences between oneself and others. We saw this line of argument clearly at work in Chapter 2, in the attempts to morally enfranchise non-human animals. However, if one is literally constituted by one's environment, then standard egoistic considerations concerning the intrinsic value of oneself will have straightforward application to the environment also. Bluntly, you should intrinsically value the environment because it is part of you, and you intrinsically value yourself.

Unfortunately, this line of reasoning is just too quick. The principal reason for this is that one's metaphysical and conceptual unity with the environment is compatible with a range of different attitudes towards it, including, notably, an attitude of manipulation or exploitation. To see this, consider essentially the same position as that of Callicott, this time developed with a slight difference in emphasis. Ruth Millikan has argued that the line between organism and environment, as it is customarily drawn, is detrimental to the proper understanding of organisms.[14] It is possible to draw a distinction here, of course; it is possible, in principle, to draw a distinction anywhere one pleases. But such a distinction will not be principled; not theoretically useful for understanding the nature and behaviour of organisms. Rather, the only principled way of differentiating organism and environment makes this distinction a matter of degree, specifically, 'the degree of control that the system as a whole has over the production and maintenance of normal structure and normal states for its various portions.'[15]

The basic idea underlying this claim runs as follows. Consider, for example, the lungs. These are an organismic subsystem and, as such, have a normal environment in the absence of which they cannot perform their normal functions. They cannot, that is, perform their function of supplying tissue with oxygen unless encased in an airtight chest cavity, are displaced periodically by certain muscles, unless next to a heart pumping blood through the relevant vessels, and unless the wider system of which they are a part (i.e. the organism) is surrounded by an atmosphere containing oxygen. The crucial point is that as we move closer to the spatial centre of any organism we find a greater proportion of systems whose normal conditions for proper functioning are maintained by some other connected system. Presence of a functioning heart, for example, is an essential condition for the normal functioning of the lungs, and there are numerous systems within the body that help maintain a functioning heart as part of their jobs. Towards the spatial periphery of the organism, however, conditions necessary for proper functioning of the organism's subsystems tend to be less under the organism's control. The proper functioning of the lungs requires that the organism be situated in an environment containing oxygen, and whether the organism is so situated depends on factors over which the organism does not always have control. It is true that if an organism finds itself in an atmosphere bereft of oxygen, it might be able to move to a more oxygen rich one, but as we move from the spatial centre of the organism, the tendency is for the method of maintaining normal conditions to become, in Millikan's words, 'less a making and more of a seeking or fitting in'.[16] At the outer limit, normal conditions for portions of the organismic system are simply there or not there, maintenance of these conditions being completely beyond the organism's control.

Millikan's position with regard to the relation between organisms and environments is essentially the same as Callicott's. Indeed, as we have seen, Callicott's case appeals to precisely the sorts of evolutionary considerations that are central to Millikan's wider account.[17] However, in developing her account, Millikan emphasises the notion of control. And this emphasis makes it clear that the metaphysical unity of organisms with environments might issue in anything but the sort of undifferentiated identification with, and quasi-mystical love of, the environment. On the contrary, even if the wider environment is literally part of us, it is a noticeably recalcitrant, intransigent, part. Something often beyond our control, therefore frightening. It is as if we had a body part, an arm, leg, or, for that matter, a penis, over which we had no control. It just hung there, limp, lifeless, inert. Or perhaps, it occasionally sprang into an unpredicted and uncontrolled life of its own. It is not clear that we would thereby *love* this part of us, simply because it is part of us. More likely, it seems, is that we would grow to resent, hate, this useless appendage. Indeed, we may redouble our efforts to subjugate and control it. The same sort of point applies to aspects of our personality and character traits that we find undesirable.

Once cannot move from metaphysical and/or conceptual unity with the environment to empathetic identification with and/or love of that environment. Many people, of course, do empathetically identify with their environment, and many more claim to. But such identification is not something which follows from our metaphysical or conceptual unity with that environment. Indeed, when we consider that the environment contains not just natural objects and properties but also man-made ones, this point should be more or less obvious. Is one to empathetically identify with the landfill one can smell when the wind comes from the north-east, simply because it is part of one's environment, hence part of oneself? In order to develop an adequate conception of environmental value, one must break down the subject/object distinction. This is true and this is important. But just as important is to break it down in the right way. One does not break it down in the right way by embracing an empathetic, quasi-mystical, merging with the environment.[18]

6
Against Humanism I: Externalism

6.1 Against the spirit of our age

We philosophers live in a humanistic age. The dominant philosophical doctrine of our time, today's intellectual *zeitgeist*, is that the world is a world structured by us; forged by the architectural propensities and proclivities of our mind. This is the Kantian turn in philosophy. Reality as it is in itself, noumenal reality, is essentially unknowable, and philosophy, accordingly, shifts from the study of being-qua-being to the study of being-qua-known. Philosophy is *first* philosophy, and first philosophy is the study of the structuring activities of the human mind. Philosophy is the philosophy of thought. This much has been the orthodoxy ever since Kant. Just think how much of philosophy in the twentieth century has been shaped by, and makes little sense without, this tenet. There are, of course, the obvious examples such as the phenomenalism prevalent in the early part of the century. Less obviously, the so called linguistic turn, which, until quite recently, dominated philosophy in the Anglo-American world, was essentially a linguistic form of Kantianism, constituted by appending to the Kantian turn one of two claims: either the structure of language determines the structure of cognition, or the structure of language mirrors the structure of cognition. Once we accept either of these, philosophy as first philosophy can proceed as philosophy of language rather than philosophy of thought. In a similar vein, much of twentieth-century philosophy of science has been exercised by the question of the so called theory impregnation of observation; the extent to which our observation, in both scientific and everyday contexts, is influenced, shaped and even determined by background theoretical principles and assumptions. And, in accordance with the Kantian spirit of our time, this claim about the content of observation becomes translated into a claim about the content of reality. It is not just observation but the reality that is observed that is laden with theory. Much of the popularity of Thomas Kuhn's *The Structure of Scientific Revolutions*, easily the best selling philosophy book of all time, almost certainly stemmed from its being interpreted in this way, an interpretation which Kuhn vainly spent the rest of his career trying to repudiate. Without too much over-simplification, one can say that the role played by the mind in Kant's worldview has been played by 'theory' in much twentieth-century philosophy of science. And if we switch from so called analytic philosophy to the allegedly antagonistic *Continental* alternative, the essentially Kantian organising vision remains. In the phenomenological tradition, Husserl, its founder, developed his own phenomenological version of transcendental idealism. And even Heidegger, who is in many respects a very unKantian thinker, and who in fact explicitly describes his position (or one of them) as *anti-humanist*, tells us that man is the lighting up place of being, the place where beings come to be. And in the structuralist and poststructuralist tradition, one could say, without an inordinate amount of oversimplification, that the role played by the mind in Kant's worldview is played by 'the text'.

Undoubtedly, much of the above is polemic. But I do not think it is that far from the truth either. And, why stop when I'm on a roll? The roots of this humanist, neo-idealist, neo-Kantian organising vision can be traced back to the very beginnings of Western philosophy. An intimate connection between reality and our knowledge of it can be clearly found in Plato. Reality, for Plato, is essentially intelligible: it is that which can be understood by reason alone, at least when reason is embodied in a suitably trained and cognitively equipped subject. The connection between reality and intelligibility is, thus, an intrinsic one. If this connection was first asseverated by the father of Western philosophy, then it was certainly extended and strengthened by the father of *modern* Western philosophy, René Descartes. Descartes made the criterion of the reality of a situation or state of affairs its representation to a knowing subject with certainty. This sort of developmental profile of the history of philosophy is sketched, with typical perspicacity, by Nietzsche in a little known passage in *Twilight of the Idols* aptly entitled 'How the Real World at last became a myth'. And, if Heidegger is to be believed, Nietzsche himself became the culmination of this history in his assertion that the world is simply will to power: 'This world is will to power and nothing else. And you yourselves are this will to power and nothing else besides.'[1] Except Nietzsche was not the culmination.

We are still culminating. The world in twentieth-century philosophy is still a fable.

The term 'humanism', as employed in this book, denotes the neo-Kantian, neo-idealist, view that the world depends for its existence, nature and properties on the human mind. The world is *ontologically* dependent upon human consciousness. A typical accompaniment to this claim of ontological dependence, an accompaniment clearly expressed in Kant's conception of philosophy as the philosophy of thought, is that the world is *epistemologically* dependent on human consciousness. If we want to know, philosophically, the world, we must study the human consciousness that constitutes it as a world, or we must study the products of this consciousness – language, theory, text, or whatever – that constitute it as such. But if the world is ontologically dependent on human consciousness, and epistemologically dependent on human consciousness, then it is very difficult to see how it can be anything other than *axiologically* dependent on this consciousness also. If the world depends for its reality on the activities of human consciousness, then it must almost certainly depend for its *worth* on human consciousness too.

Hence, the dilemma of environmental philosophy. Environmental philosophy is the branch of philosophy concerned with the worth, the value, of the environment. But the destiny of 2000 years and counting of philosophy is that this environment can have no value, or whatever value it has derives exclusively from the activities of the human mind. As we stand at the dawn of a new millennium, it is surely time to ask ourselves: *who do we think we are?* It is time to jettison Kant and his anaemic imitators. Philosophy must start again.

6.2   Externalism

In fact, philosophy has already started again. It's just that not many people in it, and nobody outside it, realises it. Now I am going to tell you a story.[2] Many of you will have heard it before, many of you will have not.

Once upon a time there was a planet. A very beautiful, unusual and philosophically convenient planet. The reason it was all these things is because it duplicated our planet earth in almost every respect. And so on this planet, which we can call *twin earth*, there was a twin you: someone who was your exact duplicate, down to the level of molecular structure. And this person led exactly the same life as you, did the same things as you, had the same experiences as you, lived in the same house as you, struggled to read the same books as you.[3] You get the idea. Earth and twin earth are almost identical, but not quite. The sole difference is this. On twin earth, there is no water. To be sure, they have rivers and oceans filled with a colourless, odourless, drinkable liquid, and the twin earthers use this liquid in the same way as we use water; they drink it, wash in it, cook with it, and so on. Indeed, since they speak twin English, they also refer to it with the word 'water'. But it is not water, since it is not a substance made up of hydrogen and oxygen. Instead, it has a complex molecular structure comprising other elements that we need not go into for the purposes of the thought experiment. The crucial point is that this substance, which *we* shall call 'retaw', but the twin earthers call 'water', is not water.

Now, suppose you and your twin both say something like 'water is wet'. Do you mean the same thing by your utterances? Well, it is difficult to see how you could possibly mean the same thing. Your utterance is about water, but your twin's utterance is about retaw which he or she happens to call 'water'. Crucially, what makes your utterance true is a fact about water, but what makes your twin's utterance true is a fact about retaw. The utterances, therefore, have different truth-conditions: different things are required of the world in order for your respective utterances to be true. But it is a virtual truism that the meaning of an utterance depends, and depends intimately, on its truth-conditions. Therefore, your utterances have different meanings. But, *ex hypothesi*, everything that is going on in your head is the same as what is going on in your twin's head. You are twins down to the level of molecular structure, and you have had exactly the same (non-intentionally specified) experiences, memories and so on. And this shows that what someone means by a statement, spoken or written, is not determined by what is going on inside his or her head. You and your twin mean different things, but what is going on inside your respective heads is the same.

The above provides the bare bones of a very famous thought experiment of Hilary Putnam. The initial reaction of many non-philosophers to it is to wonder how philosophers can possibly get away with talking about outlandish, fantastical scenarios such as this. But the twin earth scenario is, in fact, just a way of making a very simple point, a point that could be made without employing it. The point is that if we fix what is going on inside the head of an individual and vary their environment, then the meaning of what they say can vary with the changes in the environment even though what is in their head remains fixed. Or, as Putnam puts it, more sparingly, 'meanings ain't in the head'.

In fact, the same point applies not just to meaning, but also to other sorts of mental state. When you believe that water is wet, for example, do you believe the same thing as your twin when he or she has a belief that is attributed by a sentence of the same form? You do not. Your beliefs differ, and for the same reason as the divergence in your meanings. You believe something about water, but your twin believes something about an entirely different substance. Thus, what makes your belief true is not what makes your twin's belief true. Therefore, you each have a different belief. If meanings 'ain't in the head', then neither are beliefs. This view is often known as *externalism*.

The same sorts of considerations can be applied to all the so called *propositional attitudes*. Propositional attitudes are states such as believing, desiring, thinking, hoping, fearing, expecting, anticipating, and so on. What unites these states into a single category is that they all have what is known as *content*. That is, they are states whose ascription to a person involves the use of a *that*-clause, as in Jones believes *that* water is wet. Propositional attitudes, therefore, have content because they are relations to propositions which have content (or are contents), and they are connected to this content by means of the word 'that'. Propositional attitudes are not defined by a distinctive phenomenology. The phenomenologies associated with your belief that water is wet and your twin's belief that retaw is wet are identical; nonetheless the beliefs are distinct because their contents are distinct.

It might be thought that there is a clear sense in which you and your twin *do* mean and believe the same thing in the above case. That is, it might be argued that when you and your twin say 'water is wet' what you *both* mean is that the colourless, odourless, transparent . . . drinkable liquid is wet. The issues here are difficult, and this is not the place to even attempt to resolve them. Some accept that there is a component of meaning that you and your twin share. In this context, it is common to find a distinction between *narrow* and *wide* content. The idea, very roughly, is that a description such as 'the colourless, odourless, transparent . . . drinkable liquid is wet' specifies the narrow content of what you mean and believe, whereas a description such as 'water is wet' specifies the wide content. Then, the idea is that narrow content is 'in the head' even if wide content is not. Others, however, argue that by the time the various externalist arguments have been run (and there are several more that we have not considered here) the part left over that is genuinely narrow does not qualify as a species of content but, instead, must be assimilated to a form of non-conceptualised phenomenological state of some sort. So, propositional mental states do not possess two separable forms of content as such. The issues, as I say, are extremely difficult, but, happily, there is no need for us to adjudicate between them here. Instead, we can understand externalism simply as the view that there is at least a component of some mental states that is not determined by what is going on inside the heads of their subjects. Therefore, certain sorts of mental states are not, purely, 'in the head'.

6.3   The scope and limits of externalism

Externalism about mental states is easily the most significant development in Anglo-American philosophy in the second half of this century. But it has gone strangely unnoticed by those outside this relatively modest domain. For what externalism gives us, in all essentials, is an anti-Cartesian conception of the mind, consequently, an anti-Kantian, anti-idealist, anti-humanist understanding of the world. Externalism holds that the mind is penetrated by the world. Mental distinctions, distinctions between one content based mental state and another, are grounded in worldly distinctions. Differences in the mind depend on differences in the world. What is going on in the mind is, thus, individuation dependent on what is occurring in the world; the direction of individuation runs from world to mind. Accordingly, this individuation dependence is asymmetrical; the mind is individuation dependent upon the world, but *not* vice versa, since what serves to individuate a mental state cannot itself be individuated by that mental state. In other words, externalism presupposes a form of *realism* about the external world; it takes worldly facts to be fixed independently, so that these can then be used in the fixing of mental facts.[4]

Contrast this with idealism in all its forms. What is common to all forms of idealism is the idea that the mind is individuatively basic with respect to the world. Idealism, in whatever form, attempts to explain the worldly facts by reference to mental facts. And this means that the mental facts themselves must be antecedently fixed. Thus, if a Kantian, for example, tries to explain the existence of certain worldly facts in terms of the structuring activities of the mind, in terms of what Kant calls its forms and categories, then the content of these forms and categories must be taken for granted and then used to explain the occurrence of the worldly facts. And, similarly, if the poststructuralist tries to explain the appearance of certain worldly 'facts' in terms of the structuring activity of the 'text', then she must be taking the existence and content of 'text' as a free lunch. What neither the Kantian nor the poststructuralist can do is turn around and explain the content of the forms/categories or 'text' in terms of the worldly facts. In general, any form of neo-Kantian idealism must assume that the content of the mind (or its products such as theory, text, or whatever) are antecedently fixed, and then use this content to explain the nature of the world. Thus, neo-Kantian idealism presupposes what we can call an *internalist* view of the mind: the view that the contents of the mind are intrinsic to it, that they are fixed independently of the mind's relation to the world (or to anything else). And internalism, in this sense, is simply one form of our old friend: Cartesianism. The neo-Kantian, neo-idealist, *humanist* view of the world, thus ultimately requires a view of the mind as a self-contained interiority that possesses its contents intrinsically. And the significance of externalism is that it denies that the mind can be this way.

Things are no doubt less clear cut than the above sketch would suggest. Rather than being simply polar opposites, both externalism and idealism, in fact, consist in a spectrum of views, some stronger than others, some more plausible than others. And it is far from clear that all forms of externalism are incompatible with all forms of idealism.[5] Nevertheless, for our purposes at least, these complications are unimportant. What is important is the idea that externalism is a view of the mind that is essentially anti-Kantian, hence essentially anti-humanist. According to externalism, the world is not ontologically, nor epistemologically dependent on the mind. Rather, the direction of dependence runs in the opposite direction, the world is basic, the mind derivative. So, if the problems philosophy has in dealing with the value of the environment stem from its patent or latent humanism, from its according the environment, in true Kantian fashion, only a derivative ontological and epistemological status, then the potential importance of externalism should be readily apparent.

Externalism may seem a strange, counterintuitive, view. It would do, given the Kantian milieu in which we have been raised. The real problem with externalism, I shall try to show, is not that it is counterintuitive, but that it is incomplete. The problem with externalism is not that it goes too far, but that it does not go far enough. Externalism is unduly restricted in at least two ways.

Firstly, the arguments for externalism turn essentially on the *content* of mental states. That is, if they apply to a given class of mental states, then this is in virtue of the fact that those states possess content. In fact, in order for the standard externalist arguments to succeed, one must suppose not only that mental states possess content, but that they do so *essentially*. For, if not, then all the standard externalist arguments would show is that certain of the inessential features of mental states are not determined exclusively by what occurs inside the head of any given individual but depend on his or her environment. But that is very different from showing that mental states themselves are, in this sense, environmentally dependent. The upshot, of course, is that the arguments for externalism will work only for propositional attitudes – beliefs, desires, and the like – and then only if we assume that these have their contents essentially. But there are many aspects of mentality that are excluded by this restriction.

In particular, cognitive processes are excluded. A cognitive process, very roughly, is any process that is essentially involved in the solution of a cognitive task, and the notion of a cognitive task is perhaps best defined by enumeration. Thus, *perceiving* the environment, *remembering* perceived information, and *reasoning* on the basis of remembered information are all examples of pretty central cognitive tasks. Any process that is essentially involved in accomplishing these tasks counts as a cognitive process. The standard arguments for externalism do not touch cognitive processes as such. Explaining the nature of these processes is usually regarded as a matter of describing the so called *cognitive architecture* of an organism; that is, describing the various internal occurrences that allow the organism to process information in a way required for the solution of the cognitive task in hand. Indeed, the claim that these processes are purely internal ones has, until very recently, been almost constitutive of this sort of cognitive theorising. That is, it is almost universally assumed that the capacity of an organism to process the information it needs to successfully interact with the environment can be explained purely in terms of what are known as *mental representations* and operations defined over those representations. Since a mental representation is a purely internal state that carries information about the environment, the operations defined over these representations, the processes in which the representations take part, are conceived of as purely and exclusively internal operations.

Now, it is important to realise that the internality of cognitive processes is not something that is based on empirical evidence of any sort. It is not a straightforwardly theoretical claim that can be confirmed or refuted by empirical investigation. Rather, it has the status of what Wittgenstein sometimes calls a *mythology*, and, at other times, a *means of representation*. That is, the internality of cognitive processes is not something that is established on the basis of empirical evidence, but is a principle which guides and organises cognitive theorising, and, thus, the gathering and ordering of empirical evidence. The thesis of the internality of cognitive processes, as a pre-theoretical organising vision, is one more vestige of neo-Kantianism that needs to be replaced by a suitably extended version of externalism.

The second limitation that externalism fails to breach is this. The arguments that are used to support externalism fall short of fully breaking down the internalist view of mental states as located inside the head of individuals that possess them, even when we restrict our attention to propositional attitudes and forget about underlying cognitive architecture. Perhaps the best way to express the ramifications of externalist arguments is in terms of the notion of *individuation dependence* as developed by Strawson.[6] Following Strawson, we can identify four aspects to a claim of individuation dependence: linguistic, epistemological, metaphysical, and conceptual. Suppose Fs are individuation dependent on Gs. Then we can say the following: (i) reference to Fs requires prior reference to Gs, (ii) knowledge of the properties of Fs requires prior knowledge of the properties of Gs, (iii) it is of the essence of Fs that they be related to Gs, and (iv) possessing the concept of an F requires prior possession of the concept of a G.

As Colin McGinn has pointed out, externalism seems to follow this general schema quite closely.[7] Suppose the Fs are beliefs and the Gs the objects or properties environmentally related to the subject of those beliefs. Then the claim that beliefs are individuation dependent on the environment is the claim that the following four conditions hold: (i) reference to beliefs requires reference to appropriate environmental entities – there is no way of saying what beliefs a subject has except by reference to the worldly entities those beliefs are about, (ii) knowledge of beliefs requires knowledge of appropriate environmental entities – we cannot know what someone believes in a particular case without knowledge of the worldly entities her belief is about, (iii) it is of the essence of a particular belief that it be related to environmental entities – it simply is not possible for a belief to be held by a subject unless his environment or world contains the appropriate entities, and (iv) one could not master the concept of a particular belief without having mastered the concepts of the worldly entities the belief is about. These four conditions, then, express the content of the claim that beliefs (or other propositional attitudes) are individuation dependent upon environmental objects and properties.

The standard arguments for externalism, if correct, clearly show that beliefs and other propositional attitudes are individuation dependent on the environment. But do they show any more than this? The problem here is that the simple idea that beliefs are individuation dependent upon environmental entities falls far short of a convincing repudiation of a neo-Kantian, humanist, conception of the mind, and, consequently, of the world. More precisely, many, and perhaps most, dimensions of the internalist view of the mind are untouched by the claim that (some) mental states are individuation dependent upon environmental entities. Consider an apparently parallel case.[8] The property of being a planet is individuation dependent upon the property of being a star in the sense that (i) there is no way of saying what a planet is without reference to such things as stars, (ii) one cannot know what a planet is unless one knows what stars are, (iii) it is of the essence of a planet that it be related to a star, and (iv) one could not master the concept of a planet unless one had also mastered the concept of a star. Nevertheless, even though the property of being a planet is externally individuated in the sense described above, this does not mean that a token of this type, an individual planet, is *located* where its star is located. It manifestly is not. In other words, there is a clear distinction between the claim that the property of being a planet is individuated with respect to the property of being a star and the claim that each planet is located in the same place as its central star. Similarly, it seems, the fact that mental states are externally *individuated*, individuation dependent on environmental objects and properties, does not entail that they are externally *located*, that they exist, in part, in the environment itself. The standard arguments for externalism show that mental states are externally individuated, but they do not show that they are externally located. And, therefore, one crucial component of the Cartesian conception of the mind is left completely untouched by externalism. The conception of the mind as a spatially self-contained interiority remains.

The aim of the next two chapters is, in effect, to develop a generalised form of externalism, one which, hopefully, is successful in removing more of the hydra-headed Cartesian conception of the mind. This form of externalism will apply to cognitive processes and the architecture which underwrites them as well as to content bearing mental states. And this form of externalism will entail that these states and processes are both externally individuated *and* externally located. This suitably generalised version of externalism I shall call the *environmentalist model of the mind*.

7
Against Humanism II: Evolution

The task of the remainder of this book is to develop a radical approach towards understanding the value of nature. In common with the radical approaches of Callicott examined in Chapter 5, it will be argued that the question of the value of nature is ill-conceived as a question of whether this value is objective or subjective, since this makes sense only if we accept the traditional distinction between mind and world. It is this opposition that the present and the following chapters seek to break down. To this end, then, these chapters develop a generalised form of the externalism outlined in the previous chapter. Or, they develop, as I prefer to call it, an *environmentalist* model of the mind. This model dismantles the traditional mind/world distinction, and does so without appeal to, indeed in opposition to, the tired humanist idea that mind constitutes world, an idea which, I have argued, is anathema to genuine environmental thinking.

The opening chapter introduced the twin motifs, borrowed from Holderlin, of the *danger* and *that which saves*. In this book, the danger is our proclivity for seeing, understanding and conceptualising the natural environment as nothing more than a resource: as something whose value lies solely in the extent to which it can further our goals and satisfy our needs. That which saves, on the other hand, has been represented as a successful dismantling of the mind/world distinction, a dismantling that allows us to understand ourselves as properly beings-in-the-world and, on this basis, develop a satisfactory conception of the value of the natural environment. The opening chapter also hinted that the relation between the danger and that which saves might be far more intimate than polar opposition. We might be beings-in-the-world only because we are, ultimately, manipulators and exploiters of it. This is the idea that will be explored in the remaining chapters.

In what follows, the immediate and most obvious emphasis will be on the danger. I shall argue that there are deep seated reasons, reasons rooted in our biological, and consequently our cognitive, development, why we are tempted to understand the environment simply as a resource. Seeing the environment as a resource is, in one sense at least, inevitable, given our natural history. Such a conceptualisation of the environment, then, is not a peculiarly cultural product. It is not, as some have claimed, the result of unfettered industrial capitalism, nor, as others have claimed, is it the product of some sort of rampant masculinisation of thought. The roots of our exploitation of the environment do not lie in our exploitation of each other. They lie far deeper than that. As the idea of the danger is developed in the next two chapters, however, what also grows, slowly but discernibly, is the sense in which this danger can also be that which saves. Proper understanding of the danger, I shall argue, is, ultimately, what allows us to dismantle the mind/world distinction, and do so in a way that is genuinely environmentalist, rather than humanist, in character.

This chapter examines evolutionary theory. In our evolutionary development, I shall argue, lie the deep roots of both the danger and that which saves.

7.1 The concept of evolutionary cost

The juvenile sea squirt spends its days navigating its way along the ocean floor. To facilitate such navigation it possesses a rudimentary brain and nervous system. However, upon reaching maturity it fastens itself to a rock where it spends the rest of its life. And, having done this, it then proceeds to eat its own brain! Once ensconced in its rocky niche, it no longer has need of a brain. So it gets rid of it in the only way it knows how. As the neuroscientist Rodolfo Llinas observed, there is a perhaps disturbing similarity between this and the process whereby a lecturer gets tenure at a university.[1] But why does the squirt not hang on to its brain? After all, it already has it, and given the vicissitudes of life on the ocean floor, it never really knows when it might need it again.

Consider another example. Female ants can sprout wings if they happen to be nurtured as queens, but if nurtured as workers do not express this developmental capacity. Furthermore, in many species of ant the queen will use her wings only once – for her nuptial flight – and then proceed to bite or break them off at the roots.[2] Why should she do this? Indeed, why wouldn't all female ants develop wings given that they have the ability to do so? Doesn't the possession of wings count as a distinct advantage?

The answer to both these questions lies in a concept central to the arguments of this chapter: the concept of *evolutionary cost*. Any evolutionary development involves an investment of *resources*. If, for example, a lineage of creatures, such as ants, has evolved wings, then such evolution will use up a certain quantity of resources. These resources are then not available for use elsewhere in the biological economy of the organisms of that lineage. Given that any organism of this lineage has at its disposal only a finite quantity of resources, then any investment of resources in a particular evolved feature will necessarily show up as a deficit elsewhere. This seems why, for example, winged aphids are less fertile than wingless ones, even within the same species.[3] Resources that were employed in making wings are no longer available for making eggs. In evolution, there really is no such thing as a free lunch. Every evolutionary development must cost something. That is, every adaptation involves the utilisation of resources which cannot, then, be utilised for other purposes. Evolutionary development is, therefore, a constant process of balancing costs and selective advantages. For an aphid, resources that are invested in making and powering wings are resources that could have been spent on making eggs. For a sea squirt, resources that have been spent on a brain have similarly been bought at a cost that will be felt somewhere else in the creature's economy. Each point in the evolutionary development of a lineage becomes a balancing of costs and benefits. In this sense, each point in evolutionary development is a 'solution' to this balancing problem.

The answer to the questions posed by the earlier examples, then, is that wings, and brains, have their costs as well as their benefits. If the selection pressures that might lead to the development of a given feature are not sufficiently important, then the cost of developing that feature might be greater than the benefits that accrue from it.

7.2   Evolutionary cost and environmental manipulation: beaver and superbeaver

Beavers build dams. The evolutionary explanation of this dambuilding behaviour goes something like this. A dam results in the creation of a miniature lake. The presence of the lake increases the distance the beaver is able to travel by water, and this is both safer than travelling by land and easier for transporting food. If a beaver lived on a stream only, then the supply of food trees lying along the stream bank would be quickly exhausted. By building a dam, the beaver creates a large shore-line which is available for safe and easy foraging without the beaver having to make long and hazardous overland journeys. So, the building of dams became incorporated into the beaver's evolution.[4] Things, of course, could have happened differently. Instead of investing in dam-building behaviour, the evolution of the beaver might have involved investing in ways which facilitated its ability to travel overland. Suppose that in the dim and distant evolutionary past the ancestor of the beaver had started evolving in two alternate ways. The first of these ways involved adopting the dambuilding strategy and culminated in the sort of beaver with which we are familiar today. We will suppose, how-ever, that the second evolutionary strategy completely eschewed dam-building behaviour and, instead, concentrated on making the beaver stronger, quicker, and more intelligent, thus increasing its efficiency in evading predators and transporting food on the long overland journeys it was obliged to make. This sort of strategy, therefore, involved invest-ing in such features as increased muscle mass, larger brain, more power-ful legs and torso, etc. So, in this thought experiment, evolution results in two very different types of beaver: the ordinary dambuilding beaver, and the speedier, stronger, smarter *superbeaver*.

The question is: which one is fitter, the beaver or the superbeaver? First, let's do some necessary fixing of variables. The task, or rather tasks, that the beaver's ancestor had to accomplish were (i) the location/trans-port of food, and (ii) the evasion of predators. The ordinary beaver and the superbeaver have attempted to accomplish these tasks through the adoption of two alternative strategies. Since we have no *a priori* basis for supposing that one strategy is more effective than the other, let us suppose, for the sake of argument, that the ordinary beaver and the superbeaver are equally competent in the performance of these tasks. That is, statistically speaking, the ability of the ordinary beaver to transport food and evade predators by way of its strategy is equal to the ability of the superbeaver to transport food and evade predators by way of its strategy. That is, the benefits which accrue to the ordinary beaver and the superbeaver through adoption of their favoured strat-egies are the same.

Secondly, let's refine the notion of fitness. The crucial concept here is that of, what we can call, *differential fitness*. In talking of the fitness of an organism, it makes sense to speak of not only that organism's absolute fitness, but also its fitness in a particular respect. Adoption of a success-ful strategy of adaptation by an organism results in the increase of that organism's absolute fitness. Adoption of an unsuccessful adaptive strategy, on the other hand, would decrease the fitness of the organism. The change in an organism's fitness brought about by adopting a strategy of adaptation I shall call differential fitness. Since differential fitness is brought about by the adoption of a particular strategy of adaptation, differential fitness is always a function of, and is therefore indexed to, a particular adaptative strategy.

Now the question we have to ask ourselves is this. Does the fact that the ordinary beaver and the superbeaver are equally competent in the performance of their tasks entail that the differential fitness associated with the ordinary beaver strategy is equal to the differential fitness associated with the superbeaver strategy? In fact, it does not. It would entail this only if it could be shown that the *cost* of adopting the two strategies is the same. However, this does not seem to be the case. On the contrary, it would seem that the ability of the superbeaver to transport food and evade predators has been bought at a greater cost than the corresponding ability of the ordinary beaver.

When assessing the costs involved in each of these strategies, it is useful to distinguish between what we can call *implementational* and *performance* costs. The situation is very similar to the costs involved in owning a car. First there are the costs of buying the car, and then there are the day-to-day running costs. The implementational costs of a strategy are those that are involved in the setting up of the strategy. These will, principally, be the genetic costs required to develop the structures, mechanisms, and so forth that are required to implement the strategy. The performance costs, on the other hand, are the day to day running costs of the strategy, measured in terms of the energy required to perform the strategy on each occasion.

Consider, first, how the implementational costs of the two strategies compare. The implementational costs of the ordinary beaver's strategy comprise largely the genetic resources necessary for the development and maintenance of the structures which allow the ordinary beaver to adopt it. The primary structures here will be large, flat, powerful teeth and a flat tail together with the necessary surrounding musculatures. Compare these to the implementational costs of the superbeaver's strategy. Firstly, the superbeaver will require more powerful muscles for dragging its food on the long overland journeys it is obliged to make. Thus, its limbs and torso must become, pound for pound, more powerful than those of the beaver. Secondly, it must also possess the capacity to escape from the predators it will inevitably encounter on these long overland journeys. Thus, it might have to become quicker. It might also have to become more intelligent, thus creating a need for brain encephalisation. At the very least its sensory modalities would have to improve, allowing it earlier and more reliable detection of predators. This again would require encephalisation. Moreover, should encephalisation occur, then the costs would start to multiply dramatically. Encephalisation entails a larger brain, which in turn requires a larger head, which in turn leads to more weight at the front of the body. This must be balanced by added weight elsewhere which, in turn, requires stronger, i.e. larger, muscles, and so on. Given this is so, a fairly strong case can be made for the claim that the implementational costs of the superbeaver strategy are greater than those of the ordinary beaver strategy. At the very least, we can say that the implementational costs of the superbeaver strategy will not be any less than those of the ordinary beaver strategy.

It is when we turn to performance costs, however, that the disparity between the strategies really becomes evident. The performance cost of dragging food trees overland, running away from predators one encounters, and making sure one's attention is constantly tuned to their possible presence, seems to be far greater than the cost of depositing the food trees in the lake and letting the water do most of the work for you, particularly when this procedure removes the risk of predation. Of course, in terms of performance costs, the ordinary beaver's strategy requires a significant initial outlay in the form of building a dam. However, when you compare this outlay with the alternative of daily overland journeys dragging heavy food trees, then it seems clear that this outlay would soon be compensated for. The situation is very much like catching a horse and then using it. There is a significant initial outlay in the energetic costs of catching the horse, but this outlay is soon outweighed by the benefits of getting the horse to do much of your work for you. If this wasn't the case, there would have been no future in the horse-catching strategy. Or compare the initial outlay involved in building a bicycle with the benefits that arise from its completion. The latter outweigh the former. This, ultimately, is why our ancestors' strategy of building tools has been so successful.

Therefore, it is highly likely that the implementational costs of the ordinary beaver strategy are less than those of the superbeaver strategy, and the performance costs of the former are almost certainly less than those of the latter. Therefore, it seems highly likely that the overall costs of the ordinary beaver strategy are less than those of the superbeaver strategy. That is, the resources the superbeaver has invested in its ability to transport food and evade predators are greater than those the ordinary beaver has invested in the corresponding ability. Therefore, if we assume, for the sake of argument, that each strategy is equally effective in the performance of the tasks of transporting food and avoiding predators then the differential fitness associated with the ordinary beaver strategy is greater than the differential fitness associated with the superbeaver strategy. The superbeaver has invested more resources in the performance of these tasks than has the ordinary beaver. And, if this is so, the superbeaver will, thereby, have less resources to invest in the possession of other capacities. And this deficit must be felt somewhere along the line in the superbeaver's attempt to deal with its environment.

Cases similar to that of the beaver are easy to find. Not all ways of manipulating the environment, however, need be so obviously intrusive. Consider, for example, how a sponge 'manipulates' its environment in order to feed.[5] Sponges feed by filtering water, and thus require water to pass through them. This is partially achieved by way of small flagella that are capable of pumping water at a rate of one bodily volume every five seconds. However, sponges also exploit the structure of their environment to reduce the amount of pumping involved. Various adapted features – incurrent openings facing upstream, valves closing incurrent pores lateral and downstream, suction from large distal excurrent openings, and so on – make it possible for sponges to exploit ambient water currents to aid the flow of water through them. It is easy to see how such exploitation would reduce the resources the sponge needs to invest in its feeding activities. The performance costs of adopting this sort of strategy are obviously lower than those that would be involved in a pumping strategy alone, since these costs are partly met by the environment. The implementational costs are more ambiguous. However, if a sponge relied on pumping alone, and did not seek to exploit the ambient water currents, then it would need a larger or more efficient pumping apparatus in order to process water at the same rate as a sponge that did exploit the ambient currents. Developing a larger or more efficient pumping apparatus would require greater genetic investment, and this would raise the implementational costs. Therefore, the implementational costs of the strategy which eschews exploitation of the ambient water currents are at least as high as, and very probably higher than, those of the strategy which incorporates environmental exploitation. Therefore, the differential fitness associated with the exploitation of the environment strategy is greater than that associated with the non-exploitative strategy. Therefore, all other things being equal, a sponge which exploits the environment in this way would have a selective advantage over one that did not, even if the latter had a larger pumping apparatus that allowed it to process food at the same rate. The extra (genetic, energetic) resources which would inevitably be invested in the pumping apparatus would not be available for incorporation elsewhere in the sponge's biological economy.

In this case, the sponge's manipulation of the environment is a lot less intrusive than the kind of manipulation employed by the beaver. The beaver changes its environment, the sponge does not, or does so only minimally. In fact, 'exploitation' seems a better word for what the sponge does to its environment. In practice, I think, manipulation and exploitation are two sides of the same coin; each shades by degree into the other, and I shall treat them as equivalent for all practical purposes.

The environment that an organism can manipulate or exploit includes not just inanimate structures or objects but also other creatures. Sometimes this is true in the quite dramatic sense that one organism, a parasite for example, is spatially located inside another host organism. The environment of the parasite is thus made up of nothing but a living organism. And there is impressive documented evidence of the way in which parasites can survive by exploiting or manipulating their hosts. And such manipulation can include orchestration of both the structure and behaviour of host organisms.

Some parasites have a life cycle involving an intermediate, or temporary, host, from which they have to move to a definitive host, the host with which they shall remain for the rest of their lives. Such parasites often manipulate the behaviour of the intermediate host to make it more likely to be consumed by the definitive host. For example, there are two species of acanthocephalan worm, *Polymorphus paradoxus* and *Polymorphus marilis*, which both use a freshwater shrimp *Gammarus lacustris* as an intermediate host, and which both use ducks as their definitive host. The definitive host of *paradoxus*, however, is generally a mallard, which is a surface dabbling duck. *Marilis*, on the other hand, specialises in diving ducks. *Paradoxus*, then, should benefit from making its shrimps swim to the surface, while *marilis* should benefit from its shrimps avoiding the surface. Uninfected shrimps tend to avoid the light and, therefore, stay close to the lake bottom. However, when a *lacustris* becomes infected with *paradoxus* it behaves very differently. It stays close to the surface, often clinging tenaciously to surface plants. This behaviour presumably makes it vulnerable to predation by mallards and also by muskrats, which are an alternative definitive host.[6]

Paradoxus, no doubt, might have evolved in a different way. Suppose that, instead of developing a capacity to manipulate the behaviour of its intermediate host, it developed structures that enabled it to make its way to its definitive host under its own steam. In this case, *paradoxus* would have to develop structures that enable it to break out of its intermediate host, move under its own power to the region where its definitive host is to be found, evade any predators that are not suitable hosts. It seems highly likely that this alternative strategy would involve a greater investment of genetic resources than the actual strategy of *paradoxus*, a strategy that involves simply the production and secretion of a behaviour altering pheromone. Thus, the implementational costs of the alternative strategy are likely to be higher. So too are the performance costs. Breaking out of your host, fighting your way to a suitable location, avoiding and evading predators seems to involve a much higher energetic cost than getting your intermediate host organism to do all of this for you. Thus, all things being equal, the manipulative strategy of *lacustris* seems productive of greater differential fitness than non-manipulative alternatives.

The case of the acanthocephalan worm manipulating the behaviour of its host is by no means isolated or unusual. Indeed, it seems to be a fairly common evolutionary strategy. The case of the fluke or 'brain-worm' *Dicrocoelium dendriticum* provides another example. The definitive host of the fluke is an ungulate such as a sheep, and the intermediate hosts are first a snail and then an ant. The normal life-cycle calls for the ant to be accidentally eaten by the sheep. It seems that the fluke achieves this by burrowing into the brain (specifically the suboesophageal ganglion) of the ant and changing the ant's behaviour. And this change is quite noticeable. For example, whereas an uninfected ant would normally retreat into the nest when it becomes cold, infected ants climb to the top of grass stems. They then clamp their jaws into the plant and remain immobile, as if asleep. Here they are vulnerable to being eaten by the sheep. Note also that the changes in ant behaviour are not indiscriminate. The infected ant, like the uninfected one, does retreat down the grass stem to avoid death from the midday heat, since this would be bad for the parasite also. Nevertheless, the ant returns to its elevated resting position when the sun has cooled sufficiently during the course of the afternoon.[7]

Once again, the 'manipulate the ant' strategy adopted by *dendriticum* seems to be procurable at less evolutionary cost than any alternative non-manipulative strategy. A strategy of the latter sort would require investment in structures or mechanisms which not only enable the fluke to escape from the ant, but also to survive in the 'wild' until ingested by a suitable itinerant ungulate. It is fairly clear that this is a more complex, hence more difficult, strategy to execute than the actual strategy employed by *dendriticum*. Hence, it would require a greater investment of (genetic, energetic) resources than the 'manipulate the ant' strategy. Once again, the reason why the manipulative strategy involves less investment of resources than its non-manipulative alternative is because, in the former strategy but not the latter, much of the necessary work is done by the outside world, in this case another organism.

Many parasites live inside the bodies of their hosts. However, this is not true of all parasites. Indeed, some parasites may seldom come into contact with their hosts. A cuckoo is as much a parasite as *paradoxus* or *dendriticum*. This second form of parasitism, extraneous parasitism, if you like, provides us with another way in which an organism can exploit or manipulate another so as to minimise the utilisation of its own resources.

Brood parasitism provides a good example of the extraneous manipulation of one organism by another. A parent bird, such as a reed warbler, transports a large supply of food from a relatively large catchment area back to its nest. Brood parasites such as the cuckoo make a living by intercepting this flow of food. The reed warbler, of course, is not a natural altruist. The cuckoo must not only have its body inserted into the host's nest; it must also have a means of manipulating its supplier. And to this end it has evolved various features that allow it to manipulate the reed warbler's nervous system. By way of key stimuli, the cuckoo is able to exploit – to engage and employ for its own purposes – the host's machinery of parental care.[8] Most importantly, the young cuckoo, with its huge gape and loud begging call, has evolved, in a grossly exaggerated form, the stimulus which elicits the feeding response of parent birds. In fact, so much is this so that there are many records of adult birds feeding a fledged young cuckoo raised by a different host species. This demonstrates successful exploitation or manipulation by means of a *supernormal* stimulus. The cuckoo achieves the task of feeding and raising its offspring not by the sweat of its own brow, but through the manipulation of others.

Therefore, relative to the strategy of raising its young itself, this strategy of the cuckoo involves a huge reduction in performance costs. In fact, the performance costs are reduced to just about zero: after the cuckoo has laid its eggs it can completely forget about them. Some genetic resources will have been invested in the development of the supernormal stimulus (and other relevant features such as egg mimicry). But it is surely clear that whatever small increase in implementational costs there might be (and it is not even clear that there must be an increase here), this is more than offset by the virtual elimination of performance costs. Therefore, all other things being equal, the cuckoo's strategy of manipulation is selectively advantageous relative to the more traditional approach of other birds. There is nothing here, of course, which entails that cuckoos are more fit than birds that raise their young by orthodox means. It is important to distinguish differential fitness from absolute fitness. The claim is that the manipulative strategy adopted by the cuckoo produces greater differential fitness than more orthodox non-manipulative strategies. Therefore, it places the cuckoo at a differential selective advantage over other birds with respect to the task of raising offspring. However, it does not follow from this that cuckoos are at an absolute selective advantage over other birds. This would depend on a variety of factors including, most importantly, how well the cuckoo responds to and accomplishes, relative to other birds, the whole spectrum of tasks set it by evolution. How the cuckoo fares in this regard is of no relevance to the concerns of this chapter. These concerns are with the differential fitness produced by the performance of a task by way of a strategy.

The extraneous manipulation of one organism by another is not restricted to brood parasitism. The angler fish, for example, manipulates the nervous system of its prey so that it actively approaches its own doom. An angler fish sits on the sea bottom and is highly camouflaged except for a long rod projecting from the top of its head, on the end of which is the *lure*, a flexible piece of tissue which resembles some appetising morsel such as a worm. Small fish are attracted by the lure which resembles their own prey. When they approach it the angler 'plays' them down into the vicinity of its mouth, then suddenly opens it and the prey are swept into the angler's mouth by the inrush of water.[9] Instead of using massive body and tail muscles in the active pursuit of prey, the angler uses the small economic muscles controlling the rod to titillate the nervous system of the prey. Finally, it is the prey's own muscles that the angler uses to close the gap between them. The difference in performance costs between this manipulative strategy and a non-manipulative alternative is, of course, extremely significant. Instead of having to utilise massive body and tail muscles, the angler is able to capture its prey using only the small economical muscles that control its rod. The manipulative strategy, therefore, involves a clear reduction in performance costs. The implementational costs are less clear cut. Genetic resources must be invested in the rod and lure. However, there seems to be no reason why these costs would be any greater than those which would have to be invested in larger body and tail muscles should the angler have adopted a non-manipulative strategy. Therefore, once again, it seems that the manipulative strategy of the angler fish can be purchased at less evolutionary cost than non-manipulative alternatives.

Examples of this sort could be multiplied indefinitely. Similar principles emerge, for example, when we switch the focus from predatory to reproductive behaviour. Male crickets do not physically roll their female partners along the ground and into their burrows. Instead, they sit and sing, and the females come to them under their own power. From the point of view of performance costs, this strategy is much more efficient than trying to take a partner by force. In a similar vein, male canaries have not developed the means of injecting gonadotropins or oestrogens into female canaries, thus bringing them into reproductive condition. Instead they sing. The male does not have to synthesise and inject gonadotropins, he lets the female's pituitary do the work for him.[10] Evolution, presumably, might have done it differently. Male canaries might have evolved structures for the production of gonadotropins and their injection into females. But once we appreciate the concept of evolutionary cost, it is clear that this strategy, relative to the actual one, would be selectively disadvantageous: it would almost certainly be more costly than the actual one.

7.3 The barking dog principle

These examples merely scratch the surface of the ways in which organisms manipulate their environment to achieve some adaptatively specified task. It would easy to devote an entire book to the behaviour of ants in this regard, let alone the rest of the animal kingdom. But perhaps enough has been said to indicate the pervasiveness of the *manipulate the environment* strategy. The strategy is pursued by a comprehensive array of organisms: beavers, sponges, acanthocephalans, flukes, cuckoos, angler fish and canaries. And the strategy is adopted in the pursuit of a comprehensive array of ends: predation, evasion of predators, reproduction, raising of young, feeding and gaining access to food, locating suitable environments, and so on. Moreover, we have a pretty good idea of why the strategy should be so pervasive in both these senses: all things being equal, the strategy of manipulating the environment can be adopted at less evolutionary cost than strategies that do not involve such manipulation. The implementational costs of manipulative strategies are typically no greater than those of non-manipulative ones, and the performance costs are typically much less.

In short, manipulative strategies work. And they work because by adopting them you get your environment – whatever it is – to do at least some of your work for you.

Reflections of this sort thus suggest the following very general principle: for the purpose of performing a given task, don't evolve internal mechanisms which are sufficient for performing that task when it is possible to perform it by way of manipulation of the environment.[11] When an organism develops a strategy of adaptation based on manipulation of the environment, the organism must develop *some* internal, or bodily, structures. The dambuilding strategy of the beaver, for example, requires the development of large flat teeth, flat tail, and so on. However, the implementational investment here is typically no greater than the investment required for a non-manipulative strategy, and the resulting performance costs are typically far less. Thus, from the standpoint of evolutionary fitness, it is in general more effective to accomplish a given task, or solve a particular problem, through the development of capacities to manipulate the environment rather than through the development of purely internal structures or mechanisms.

Ultimately, what justifies our confidence in this principle, I think, is not evolutionary theory but basic *thermodynamics*. In any system constituted by a fixed amount of energy, if that system is required to do work, then the more of this work that can be done by something external to the system, the more energy the system will have left over after the work is done. Or, more informally, if you can get someone or something else to do your work for you, then the less work you will have to do yourself. The imagined biological case described above provides one application of this basic thermodynamic principle.

There is an old adage that seems to capture this idea quite nicely: *why keep a dog if you are going to bark yourself.* Or, closer to the present point, if you do have a dog, then you don't have to bark yourself. And getting your dog to do your barking for you will save you considerable investment of resources (i.e. energy). We can call this the *barking dog principle*, and give it a more precise formulation as follows:

> If it is necessary for an organism to be able to perform a given adaptive task T, then it is, from an evolutionary point of view, *dis*advantageous for that organism to develop internal mechanisms sufficient for the performance of T when it is possible for the organism to perform T by way of a combination of internal mechanisms and manipulation of the environment.

That is, given the 'option' of two evolutionary strategies, (1) developing internal mechanisms which by themselves give an organism the ability to perform a given task, and (2) developing internal mechanisms which, when combined with a certain type of environmental manipulation, give an organism the ability to perform the task, then the second strategy is, from an evolutionary point of view, more advantageous than the first. That is, the differential fitness associated with adoption of the latter strategy is typically greater than that associated with adoption of the first. And this is because, in general, the second type of strategy can be adopted at less evolutionary cost than strategies of the first sort.

7.4   Objections and replies

One fairly obvious objection centres around the rather loaded terminology with which I have chosen to describe adaptative strategies of organisms. The central claim of this chapter has been that the manipulation of the environment is, in an important sense, part of the natural history of organisms, that they have evolved in such a way that the environment simply *is* a resource for them. And where it is appropriate to speak of their understanding of the environment, the environment will be understood by them as a resource. However, it might be objected that this conclusion derives not from the facts of the case, but from the way I have chosen to conceptualise those facts. Concepts such as *cost*, *biological economy, resources, investment*, and the like, automatically prejudice the issue by making the interaction of organisms with their environment essentially an economic one. The view of the environment as a resource, then, is derived not from a dispassionate account of the ways in which organisms interact with the environment but, rather, from the representational devices in terms of which I have chosen to represent such interaction.

It would be no defence to point out that the conceptual devices employed in this chapter are, far from being idiosyncratic, completely orthodox devices that abound in evolutionary literature. What is more important, however, is that these devices can easily be jettisoned. What is crucial to evolutionary theory is the idea of a finite quantity of energy and a resulting zero-sum competition between organisms for this energy. The notion of a resource can ultimately be understood in terms of energy (specifically trophic energy). And evolutionary adaptations can be understood as strategies for controlling energy. Similarly, the notion of evolutionary cost can be understood as the energy that is required to be put into a strategy in order to get it to work. It is these concepts that are employed in game theoretical attempts to mathematically model evolutionary processes. The whole of this chapter could easily have been written in these terms. That it was not derives simply from considerations relating to ease of exposition. It does not derive from any illegitimately economic conceptualisation of evolutionary processes.

One important objection to the line of argument developed above might run as follows. How do we know, in specific cases, whether it is in fact more costly to develop an internal mechanism in order to accomplish a given adaptive task than to find a way to exploit or manipulate a feature of the environment? The danger is that, while we might identify some of the costs that enable us to defend the story we want to tell, we might also overlook others that point in the opposite direction. We might call this the *hidden costs* objection.

I think there are two replies that can be levelled against this objection. Firstly, the question of the costs versus benefits involved in any adaptive strategy is an empirical question that can be answered only by a detailed empirical investigation. It would be *hubris* of the highest order to present the arguments of this chapter in such a way. However, the arguments of this chapter, if correct, do create a *presumption* in favour of the *manipulate the environment* strategy. Firstly, they show that the strategy is widespread. And then, secondly, they present an explanation for why this should be so that is intuitively very powerful. The intuition is simply this. If you have a certain amount of work to do, then if you can get someone or something else to do part of this work for you, then you have less work to do yourself, provided that the work you must put in to getting this someone or something else to work for you is less than the work they thereby do. This claim is a truism. And because of this the burden of proof is surely on someone who denies that it applies in a particular case. It is not enough, therefore, to merely cite the possibility of hidden costs, costs hitherto unsuspected, one must also give some sort of reason for positively believing there are such costs, and, therefore, also some indication of what these costs are.

The above reply is related to, and reinforced by, a second point. There is something fundamentally unsound, at the methodological level at least, with the hidden costs objection's appeal to hidden costs. The reason is that hidden costs can cut both ways. Suppose you attach your car to your neighbour's car, thus getting a *free ride*. Your performance costs are thus substantially reduced. It could be objected, of course, that hooking up your car in this way involves a certain (implementational) investment – towing hitches, welding, etc. And the obvious reply here is that the extra implementational costs are outweighed by the reduction in performance costs. However, it could be objected: 'How do you know there aren't any hidden costs, ones that you haven't even thought of, in the hooking up of your car to your neighbour's?' This sort of introduction of hidden costs would be methodologically dubious. The answer to the question, in brief, is that we obviously don't know there are no hidden costs (*ex hypothesi*, if they exist they are hidden). However, we don't have any reason for supposing that there are such costs. And, crucially, we have just as much reason to suppose there are hidden costs we have not thought of with *not* hooking up our car; i.e. hidden costs in the normal running of our car. This is the reason why the raising of the possibility of hidden costs is methodologically suspect: there is as much reason to suppose that non-manipulative strategies have hidden costs as to suppose that manipulative strategies have them. In other words, the possibility of hidden costs cuts both ways and, therefore, cannot be used as an objection specifically to manipulative strategies. The methodological moral seems to be this. Hidden costs are possible, but, since they cut both ways, it is better, in arguments of this sort, to bracket them off and focus on manifest costs. Given that this is so, we have an intuitively strong reason for supposing that manipulative strategies can be adopted at less cost than non-manipulative ones. With manipulative strategies some of the required work is done by the environment. The organism who adopts this strategy, then, has correspondingly less work to do.

Another possible objection focuses on the longer term benefits of internalist, non-manipulative, strategies. Why, in the long run, can't non-manipulative strategies be more beneficial, having a pay-off which manipulative strategies do not have? This is, in effect, the converse of the hidden costs objection. While the previous objection appealed to the possibility of hidden costs associated with a manipulative strategy, the present one appeals to the possibility of hidden benefits associated with non-manipulative strategies. So let's call it the *hidden benefits objection*. This objection acknowledges that, in the short term, manipulative strategies can be adopted at less cost, but claims that the long-term benefits that accrue to organisms that have adopted non-manipulative strategies can outweigh this reduction in cost. Therefore, in the long run, manipulative strategies could be less selectively advantageous than non-manipulative ones even though they can be implemented at less cost.

One might, therefore, be tempted to dismiss this objection as methodologically suspect in the same sort of way as was the appeal to hidden costs. However, the present objection, I think, has more to recommend it to the extent that it can be supported by the following sorts of intuitions. Suppose we are faced with a certain task, say lifting a weight. A manipulative strategy for performing this task might involve getting somebody else to lift the weight for you. A non-manipulative strategy might involve engaging in a programme of bodybuilding until you are capable of lifting the weight yourself. In the short term, for the accomplishment of the task of lifting that particular weight, the manipulative strategy involves a substantial reduction in performance costs (provided that it is not too difficult to get the other person to do the work for you). However, it could be argued, the non-manipulative strategy has ramifications not only for the present task but also for your future abilities. After engaging in the bodybuilding programme, you are now able to perform many tasks of which you were not physically capable before. Thus, it could be argued, in the long run, implementing the non-manipulative strategy would be more beneficial. More generally, it might be thought that while non-manipulative strategies involve greater investment of costs, nonetheless this will be outweighed by the fact that the internal structures developed will endow one with the capacity to perform many other sorts of tasks, tasks beyond the original one for which the structures were initially evolved.

There are, however, two major problems with this line of argument. Firstly, the same sort of argument could be made on behalf of the external structures employed by the manipulative approach. They too can perform tasks beyond that for which they were originally developed. The dam that the beaver builds to facilitate food transport and predator evasion also serves as a home. Secondly, and even more importantly, the appeal to hidden benefits misunderstands the nature of evolution. In particular, evolution does not act *in the long run*. That is, evolutionary pressures do not select for long-term advantage over short-term disadvantage. Quite the contrary. Evolution crosses each bridge as it comes to it, and has no conception of the possibility of further bridges. Now, while it might be an extremely good thing for an organism to develop internal structures to perform a given task since these structures might give it the capacity to perform other important tasks also, the fact that it is a good thing cannot possibly be recognised by evolution. There may be good reasons to develop internal structures in this context, but these are not, and cannot be, good evolutionary reasons. Evolution does not act *in the long run*.

7.5   The danger as natural history

This chapter has tried to show that manipulation or exploitation of the environment is a widespread evolutionary strategy. It has also tried to show why this should be so, indeed, must be so. Dealing with the environment as a resource is not an optional extra: for most creatures it is a basic biological necessity. Most creatures have evolved in such a way that dealing with the environment as something to be manipulated or exploited is part of what they are. It is part of their natural history. Evolutionary constraints mean that, for most creatures if not all, the environment must be dealt with as a resource.

It would be a serious mistake to think that human beings are somehow exempt from these constraints. Human beings, I shall argue, no less than any other organism, have evolved to deal with the environment as a resource. Moreover, for organisms that are capable of conceptualising or understanding the world, this mode of understanding will, to a significant extent, reflect the way in which the organism has evolved to pre-conceptually deal with its environment. This is the theme to be developed in the next chapter. There, the focus switches to the development of human beings, and, in particular, the development of those capacities that have traditionally been thought to distinguish humans from the rest of the natural world: cognitive capacities. I shall argue that these have developed in a way that essentially involves manipulation or exploitation of structures in the environment. Understanding the environment as a resource, as something to be manipulated and exploited is, then, for us humans, not an optional extra: it is a part of our natural history, and essentially bound up with those capacities traditionally thought of as distinctively human.[12]

8
Against Humanism III: Cognition

In the previous chapter, it was argued that manipulation or exploitation of the environment is a widespread evolutionary strategy. It also tried to show why this should be so. Dealing with the environment as a resource, as something to be manipulated and exploited in the accomplishment of one's day to day tasks, is not an optional extra: for most creatures it is a basic biological necessity. Since evolutionary strategies that involve manipulation or exploitation of environmental structures can, in general, be adopted at less evolutionary cost than strategies that do not, any organism that adopts a non-manipulative strategy in the performance of a given task is, at least with respect to that task, differentially less fit than an organism that adopts a manipulative one. The former creature, then, risks being outcompeted for the possession of any jointly coveted environmental niche. Evolutionary constraints, therefore, mean that for most creatures, if not all, the environment must be dealt with as a resource.

It would be a serious mistake to think that human beings are somehow exempt from these constraints. Human beings, in common with every other organism, have evolved to cope with certain environmental pressures and, therefore, almost certainly have evolved to deal with the environment as a resource. Moreover, human beings are capable of conceptualising or understanding the world, and this mode of understanding, I shall try to show, inevitably reflects the pre-conceptual resource-based understanding bequeathed us by our evolutionary history. In this chapter, then, the focus shifts from general evolutionary considerations to the specific factors involved in the development of human cognitive capacities; those capacities that allow us to understand and conceptualise the world. I shall argue that the very development of cognitive capacities essentially involves manipulation or exploitation of the environment. Understanding the environment as a resource, as something to be manipulated and exploited is, then, for us humans, not an optional cognitive extra, any more than an optional biological one.

8.1   The irrelevance of exaptation

All that is needed to apply the principles identified in the previous chapter to cognitive processes – processes such as perceiving, remembering, and reasoning – is the assumption that such processes are the products of evolution. More precisely, whatever internal structures and mechanisms we possess which allow us to accomplish cognitive tasks, these structures and mechanisms have evolved through natural selection. And they have evolved at all because in our evolutionary past we were faced with certain cognitive tasks whose solution was conducive to our survival. And all this is surely undeniable. Given that this is so, then the *barking dog principle* could apply to us, and to the strategies we have adopted in order to accomplish these cognitive tasks. Not only *could* they apply to us, if they did then it means that we would have adopted the best – that is, the most efficient – type of evolutionary strategy for the development of our cognitive capacities.

The claim that cognitive processes are the products of evolution might, however, be challenged. The challenge I have in mind, here, stems not from any naive creationist nonsense, but from a much more respectable source. Gould and Vrba, while accepting that we are naturally evolved entities, deny that the structures and mechanisms that allow us to cognise are evolutionary products in the relevant sense. They write:

> The brain, though undoubtedly built by natural selection for some complex set of functions, can, as a result of its intricate structure, work in an unlimited number of ways quite unrelated to the selective pressure that constructed it ... Current utility carries no automatic implication about historical origin. Most of what the brain does now to enhance our survival lies in the domain of exaptation – and does not allow us to make hypotheses about the selective paths of human history.[1]

An adaptation, according to Gould and Vrba, is 'any feature that promotes fitness and was built by selection for its current role'. Exaptations, on the other hand, are 'characters ... evolved for other uses (or no function at all), and later coopted for their current role.' It might be argued, then, that the structures and mechanisms responsible for human cognition are exaptations, rather than adaptations. And, if this were so, it might be thought that the arguments developed in the previous chapter would be undermined.

Happily, however, the arguments of the previous chapter do not presuppose that the structures and mechanisms presently responsible for cognition evolved for that purpose. The distinction between adaptations and exaptations is simply irrelevant to these arguments. The claim is that when a structure or mechanism takes on the role of producing a cognitive process, *whether it was originally evolved for this role or not*, then it is best, from the point of view of evolutionary cost, for that mechanism to fulfil its role in conjunction with manipulation of the environment. One must never forget that evolution acts not just to *produce* structures and mechanisms, but also to *maintain* them in existence once they have been produced. A structure that is maintained because of its role in underwriting certain cognitive processes, even if it was not originally developed for this role, is an evolutionary product no less than a structure which was originally evolved for the role. Both adaptations and exaptations are products of evolution, and considerations of evolutionary cost apply equally to both of them. Whether a feature has originally evolved for its role in underwriting a cognitive process, or whether it has been coopted for this role, it is better, from the point of view of evolutionary cost, for it to fulfil this role in conjunction with manipulation of relevant structures in the environment.

The evolutionary arguments developed in the previous chapter *could*, then, apply to the development of our cognitive capacities. And *if* they did apply, we would have every reason to expect that cognitive processes would be made up of a combination of internal processes plus manipulation or exploitation of environmental structures. However, we cannot simply assume that the truth of the evolutionary argument automatically entails the claim that cognitive processes involve manipulation of environmental structures. Because, to do this we would have to assume that the development of our cognitive capacities has followed the most efficient evolutionary path. This sort of assumption is often described as *Panglossian*, after Dr Pangloss, a character in Voltaire's *Candide*, who claimed, in Leibnizian fashion, that we live in the best of all possible worlds. Perhaps the development of our cognitive capacities did not follow the most efficient path after all. And perhaps, therefore, they do not involve manipulation of environmental structures.

What we need to get around this problem is an *independent* argument for the claim that cognitive processes involve manipulation of environmental structures. That is, what we need is an argument for the environmental character of cognitive processes that does not depend on the arguments developed in the previous chapter. And it is this sort of independent argument that the present chapter seeks to provide. However, proper treatment of this issue is something that would require at least an entire book, and such treatment I have attempted elsewhere.[2] The following, then, is simply an outline, a *sketch* of an environmentalist model of cognition.

8.2   An environmentalist model of perception

Most definitions of the notion of a cognitive process begin by defining the notion of a cognitive *task*, usually by enumeration. Thus, the concept of a cognitive task includes such things as perceiving the world, remembering perceived information, drawing logical inferences about unperceived states of the world on the basis of such information, and so on. A cognitive process is then defined as (i) one that aids in the accomplishing of a cognitive task, and (ii) that does so by processing information. And the processing of information is achieved by way of the manipulation or transformation of information bearing structures. Condition (ii) is required to exclude processes that are necessary for accomplishing cognitive tasks but which do not themselves count as cognitive. For example, one cannot accomplish any cognitive tasks if one is dead, but respiration does not thereby count as a cognitive process since it does not involve manipulation of information bearing structures.[3]

On standard accounts of cognition, information processing is seen as consisting exclusively in the manipulation and transformation of *internal* information bearing structures. These internal structures are often called *mental representations*. The task of the present chapter, however, is to develop what I shall refer to as an *environmentalist* model of cognition. This is a model of cognition according to which cognitive processes are essentially made up not just of the manipulation and transformation of mental representations, but also of the manipulation and transformation of *environmental* structures. Cognition, on this view, then, is not, as is commonly thought, just an internal process of organisms. It is one which straddles both internal and external processes. Internal processes such as the transformation of neural structures are, no doubt, centrally and essentially involved. But so too are processes of

manipulating, exploiting, and transforming external structures. Manipulation and exploitation of the environment is, thus, an essential feature of the environmentalist model. This section focuses on the most
basic type of cognitive process: perception. More precisely, the focus
here will be on *visual* perception. Later sections extend the model developed here to other processes such as remembering and reasoning.

Let us begin by considering an influential account of visual perception
that we can take, for our purposes at least, to be a paradigmatic repres-
entative of the Cartesian, internalist, tradition. This account is that
developed by David Marr.[4] According to Marr, the input for visual
processing consists of a *retinal image*, where this consists essentially in
light energy distributed over an array of locations in the retina, this
distribution being created by the way light is reflected by the physical
structures viewed by the observer. The goal of early visual processing
operations is to create from the retinal image a description of the
shapes or surfaces of objects, and their orientations to, and distances
from, the observer. The first stage in this early visual processing consists
in the construction of what Marr calls the *primal sketch*. This records the
changes of light intensity in the image and makes some of the more
global image structures explicit. Construction of the primal sketch has
two stages. Firstly, there is the construction of the *raw* primal sketch; a
representation of the pattern of light distributed over the retina in
which information about the edges and textures of objects is made
explicit. This information is expressed in a set of statements about the
edges and blobs present in the image, their locations, orientations, and
so on. Secondly, the application of various grouping principles results in
the identification of various larger structures – boundaries, regions, etc.
This more refined representation is known as the *full* primal sketch.
Further processing (specifically analyses of depth, motion, shading
etc.) result in what Marr calls the $2\frac{1}{2}$D sketch. This is the culmination
of early visual processing, and describes the layout of structures in the
world from the perspective of the observer. A further and equally essential aspect of vision is recognition of objects. In order to recognise the
object to which a given shape corresponds, a third representational level
is needed; a level centred on the object rather than the observer. This
level consists in what Marr calls 3D model representations. It is at this
stage that a stored set of object descriptions is utilised. 3D model representations are the culmination of visual processing.

For our purposes, the details of each processing operation are not
important. What is important is the way in which Marr factors off
the contribution of the world from the contribution of the mind. The

contribution of the world consists purely in providing the light energy distributed across the retina. That is, the contribution of the world consists purely in providing the necessary material for visual *sensation*. This distribution of energy is informationally impoverished, at least relative to the information contained in the final perception. Therefore, the information contained in the retinal image must be *processed*. This is where the contribution of the mind takes over. Initial processing of the retinal image results in the raw primal sketch, more processing yields the full primal sketch, yet more processing is required to yield the $2\frac{1}{2}$D sketch, and yet more processing is required in order to obtain a 3D model representation. At each stage, this processing is conceived of as an essentially internal operation: a contribution made by the perceiving subject without which perception would be impossible. Marr's theory, in this way, provides a theoretical articulation of a picture inherited, in all essentials, from Descartes. It is this picture of the mind and its operations that an environmentalist model attempts to subvert.

The starting point for any environmentalist model of visual perception is what James Gibson calls the *optic array*.[5] Light from the sun fills the air, and the environment is, as a result, filled with rays of light travelling between the surfaces of objects. At any point, light will converge from all directions. Therefore, at each point in the environment there is what can be regarded as a densely nested set of solid visual angles composed of inhomogeneities in the intensity of light. Thus, we can imagine the observer, at least for the present, as a point surrounded by a sphere which is divided into tiny solid angles. Each of these contain light reflected from different surfaces, and the light contained in each segment will differ from that of other segments in terms of its average intensity and distribution of wavelengths. This spatial pattern of light is the optic array.

For our purposes, what is crucial is that the optic array is an *external information bearing structure*. It is external in the quite obvious sense that it exists outside the skin of perceiving organisms, and is in no way dependent on such organisms for its existence. It also carries information about the environment in virtue of the fact that the structure of the optic array is determined by the nature and position of the objects (or, rather, their surfaces) from which it has been reflected. The boundaries between each segment of the optic array, therefore, provide information about the structure of the environment. At a finer level of detail, each segment will, in turn, be subdivided in a way determined by the texture of the surface from which the light is reflected. Therefore, at this level also the optic array can carry information about further properties of objects and terrain. Not only is the optic array external to the perceiving organism, so too is the information it carries. Information, in this context, is simply nomological dependence. The structure of the optic array depends, in a lawlike way, on the structure of the environment. And, in virtue of this, the optic array carries information about the structure of the environment.

The optic array is, thus, a source of information for any organism equipped to take advantage of it. But the optic array does not simply impinge on passive observers. Rather, the living organism will actively sample the optic array. The perceiving organism obtains information from the optic array by actively exploring it, and thereby actively appropriating the contained information. One way of doing this is by moving, and thus transforming the ambient optic array. By effecting such transformations, perceiving organisms identify and appropriate what is known as the *invariant* information contained in the optic array. Invariant information is information that can be made available only through the transformation of one optic array into another. Consider, for example, what is known as the *horizon ratio relation*.[6] The horizon intersects an object at a particular height. All objects of the same height, whatever their distance, are cut by the horizon in the same ratio. The horizon ratio relation provides an example of the invariant information contained in the optic array. And an organism can detect such invariant information only by moving and, hence, effecting transformations in the ambient array.

What is crucial here is that (i) the optic array, a structure external to the perceiving organism, is a locus of information, and (ii) an organism can appropriate or make this information available to itself through acting upon this structure and effecting transformations in it. In this way, the perceiving organism uses or employs an external structure to make available to itself information which it can then use to deal with its environment. On standard definitions of the notion of a cognitive process, this acting upon an external structure counts as one. It is a process which aids in the accomplishing of a cognitive task, and which does so by manipulating or transforming an information bearing structure. It is just that, in this case, the information bearing structure happens to be external. Indeed, manipulating the array to make available its contained information is, in effect, a form of *information processing*. If we want to think of perception in terms of information processing – and this is a standard view – then there is no good reason for insisting that the relevant information processing occurs only inside the skin of perceiving organisms. The external optic array is an information bearing structure, and the organism, by acting upon it, manipulates and transforms this information bearing structure. Again, according to standard definitions, this is precisely what information processing is supposed to be.

Of course, one could always stipulate that the concept of information processing is to be restricted to processes occurring inside the skins of organisms. One could make a similar stipulation for the concept of a cognitive process. One can stipulate anything one likes. However, given that information is embodied in structures external to organisms, and given that an organism can manipulate and transform these structures in order to appropriate this information, it is difficult to see the theoretical salience of the restriction. When the goal is to make available to oneself information relevant to the accomplishing of a given cognitive task, and when the method is to effect transformations in relevant information bearing structures, the theoretical relevance of the distinction between inside and outside collapses.

If, relative to the concept of information processing, there is no theoretical salience to the distinction between inside and outside, considerations of simplicity apparently oblige us to say that when an organism acts upon an optic array and, by effecting transformations in this array, appropriates, or makes available to itself, the information contained therein, the organism is, in effect, engaging in a form of information processing. And, then, given the standard view that cognition is information processing performed in the accomplishing, or attempted accomplishing, of a cognitive task, this gives us the more general principle: *in certain circumstances, acting upon external structures is a form of cognition.* And this is the central or constitutive claim of what I have called the environmentalist model of cognition. This is not to claim, of course, that *no* cognitive processes occur purely inside the skin of organisms. It claims that *not all* cognitive processes occur inside the skin of organisms. Information processing and hence cognition, occur not just in the head; they occur also in the world.

The claim that cognitive process are not purely internal to organisms is a seemingly radical one. Yet it can be established from completely uncontroversial premises. Firstly, that information can be carried by structures that are external to cognising organisms. Secondly, that such organisms can appropriate at least some of this information by acting upon such structures. Once we accept these premises, and certain standard definitions of the concept of a cognitive process, the externality or environmental character of at least some cognitive processes is, I think, inescapable.

8.3   Extending the model: memory

Cole, Hood and McDermott describe a case where a child is seeking an ingredient for baking a cake.[7] The child does not need to remember exactly where the ingredient is located. Instead, the child simply goes to the shelf and works her way along it until the ingredient is found. In this case, part of the environment (the shelf) goes proxy for a detailed internal memory store.

The idea of using environmental cues to aid memory is, of course, well known. Tying a knot in one's handkerchief, writing down a shopping list, making calendars, asking someone to remind you to do something, leaving something at a special place where it will be encountered at the time it needs to be remembered, and so on, all constitute a small subset of the number of external memory aids people employ. These are typically viewed precisely as external *aids* to memory; as extrinsic devices that might help trigger the real process of remembering, itself regarded as a purely internal operation. It is this dichotomy between the internal process of remembering and external aids to remembering that an environmentalist model of memory breaks down.

Examples of the Cole, Hood and McDermott variety can be extended indefinitely. Suppose I want to find a book at the library. I have seen the book before and think it might be useful to what I am now doing. But I cannot remember what it is called or who the author is. Moreover, I do not remember exactly where it is in the library, but I do remember the floor, and I also remember that it is on a shelf with a peculiar distinguishing feature, say a distinctive colour. Therefore, I go to the correct floor, look for the shelf with the distinguishing feature, and then work my way along the shelf until I find the book. In this case, the library floor and the shelf together seem to go proxy for a complex internal memory store. Similar remarks would apply to the process of navigating one's way around the environment by way of remembered landmarks (I can't remember exactly where X lives, but I know that if I turn right by the lake, go on as far as the pub and then take a left . . .). In this case, the environment also seems to stand in for a detailed memory store.[8]

In the above sorts of cases, just as with perception, the environment contains structures that carry information. This information is relevant to the accomplishing of a cognitive task, and it can be appropriated by organisms that are capable of acting upon the structures in the right sort of way. The information processing operations required to accomplish certain sorts of memory task, therefore, consist not just in the manipulation and transformation of internal information bearing structures (i.e. mental representations) but also in the manipulation and transformation of external information bearing structures. Once again, the distinction between the information that gets processed in the head and that which gets processed outside the head is of little relevance for understanding a cognitive process such as memory.

The above examples all concern the present employment of memory capacities. But the role of environmental manipulation is, perhaps, even more important in the *development* of such capacities. Consider how the existence of environmental information bearing structures will influence the development of our capacity to remember. Perhaps the most important environmental structures in this regard are visuographic systems of representation, of which writing is the most obvious example. Consider an early form of such representation.

One of the more ancient forms of visuographic representation are known as *kvinus* ('knots' in Peruvian). These are a system of knots, and were used in ancient Peru, China, Japan, and other parts of the world. They are conventional external representations which require appropriate knowledge on the part of those tying the knots. These *kvinus* were used in Peru for recording chronicles, for the transmission of instructions to remote provinces, and for the recording of detailed information about the state of the army, taxes, etc. The *Chudi* tribe of Peru had a special officer assigned to the task of tying and interpreting *kvinus*. Originally such officers were rarely able to read other people's *kvinus*, unless they were accompanied by oral comment. However, with time the system was improved and standardised to such an extent that it could be used to record all the major matters of state: census, taxes, laws, events, etc.

Kvinus are obviously a fairly basic form of visuographic representation. Nevertheless, as Luria and Vygotsky point out (in a work that is both dated and flawed, but still classic), invention of *kvinus* will have had a profound effect on the development of strategies of remembering.[9] To see this effect, compare the memory of, for example, the envoy transmitting word for word the lengthy message of his tribal chief with the memory of the Peruvian *kvinu* officer. The envoy has to remember not simply the content of the message. Much more difficult, he has to remember the precise sequence of words uttered by his chief. The *kvinu* officer, on the other hand, does not have to remember the particular information contained in the knot he has tied, he simply has to remember the *code* that will allow him to extract this information. The envoy must rely on what is known as *episodic* memory: memory for concrete, specific, detailed events. The *kvinu* officer's reliance on this sort of memory is much less; he must employ it only in the learning of the code. Then he is able to *tap in* to the information contained in the knot, and a potentially unlimited quantity of information becomes available to him without further employment of his episodic memory.

Once an environmental information store becomes available, it is easy to see how memory is going to develop. Episodic memory becomes a lot less important, restricted to the learning of the code necessary for the appropriation of the information that is environmentally embodied. Luria and Vygotsky conjecture that this is why, during the process of cultural development, the outstanding episodic memory of members of less developed cultures tends to wither away. Members of non-literate cultures tend to rely far more heavily on episodic memory, and consequently their episodic memory capacity is much greater than that of the average member of literate cultures. With the development of an external information store of the sort the written word provides, this memory gradually becomes much less important and, as a result, vestigialises. A similar pattern can be observed in the development of children as they acquire linguistic skills. Until they reach a certain stage of development, children seem to rely a lot more heavily on episodic strategies of memory storage. Their episodic memories are, consequently, generally much more powerful than those of literate adults. Photographic memory is, among literate cultures, found mainly in children; in adults it is the rare exception. The cultural evolution of memory is, in this sense, also *involution*; constituted not only by an improvement in certain areas, but also by retrograde processes whereby old forms shrink and wither away. With the development of external forms of representation, internal development now becomes external.

As Merlin Donald points out, the development of external forms of memory storage constitutes a hardware change in the cognitive architecture of human memory, albeit a non-biological hardware change.[10] In this context, Donald introduces a useful analogy, that of *networking*. For any computer, it is possible to identify both hardware and software properties of that computer. The processing capacities of a given computer can be described in terms of its hardware features (such as memory size, central processing and speed) and its software features (such as its operating system and available programming support). The distinction is also commonly applied to humans: a hardware description is usually thought of as a description at the level of neurophysiology; a software description might describe the skills, language and knowledge carried by the individual.

However, as Donald points out, both the software *and* the hardware features of an individual computer will change if it is embedded in a network of computers. In any specification of what a computer can do, the features of the network in which it is embedded must be taken into account. Networking, that is, essentially involves a structural change. Once it is part of a network, the computer can delegate to other parts of the system computations that exceed its own internal capacity. It can also assign priorities within the larger system of which it has become part. Its outputs can be stored anywhere in the network, that is, in external information stores. In a true network, the resources of the system are shared, and the system functions as a unit larger than that of its individual components.

Another relevant analogy (also due to Donald) might be the addition of magnetic tape or disc storage to a computer. The random-access-memory of a computer is somewhat analogous to biological memory in that it depends for its existence on the intrinsic properties of the machine. One way of expanding the memory capacity of a computer is to bolster its RAM. This, however, could not go on indefinitely, and would be extremely costly. Similarly, an expansion of biological memory capacities would eventually be stymied by considerations of evolutionary cost, energy requirements, etc. (see Chapter 7). The other way of increasing the memory capacity of a computer is by way of hard external storage, and this is far more economical and, indeed, flexible than the strategy of bolstering its RAM. In essentially the same way, human memory capacity can be increased, in a similarly economical and flexible manner, by the development of external representational systems.

The invention of external representational systems such as writing should be regarded as, in the first instance, a hardware rather than a software change. Such systems constitute a collective memory hardware, and this is as real as any external memory device for a computer. Individuals in possession of reading, writing and other relevant skills have access to the 'code' that allows them to plug into this external network. They thus become somewhat like computers with networking capabilities; they are equipped to interface, to plug into whatever network becomes available. And once plugged in, their memory capacities are determined both by the network and by their own biological endowments. Those who possess the 'code', the means of access to the environmental structures, share a common memory system. And as the content of this system extends far beyond the scope of any single individual, they system becomes by far the most significant component in the memory of individuals. For humans engaged in an information retrieval (i.e. memory) task, the major locus of stored information is located 'out there' in the external representational system. It is not stored biologically, but environmentally. Biological retrieval strategies are largely limited to storage and retrieval of the relevant code that yields access to this environmental store of information, and not with the storage and retrieval of a great deal of specific information.

Therefore, in the case of memory, just as perception, it seems plausible to suppose that the information processing relevant to a given memory task can occur when an organism uses, or acts upon, a physical structure in its environment and, in so doing, appropriates or makes available to itself information relevant to that task. That is, an organism can process information relevant to memory task T through acting upon, and thus effecting transformations in, physical structures in its environment. Once again, this is not to say that all remembering involves acting upon environmental structures, only that some of it does. In the case of memory, as in perception, the distinction between the transformation of internal information bearing structures and transformation of external information bearing structures is not a theoretically salient distinction. Acting upon environmental information bearing structures is a form of information processing, hence a form of cognition. And this is precisely what is claimed by the environmentalist model of cognition.

8.4 Reasoning

In developing his dualistic account of human beings, one of the primary motivating factors for Descartes was *rationality*. Minds are essentially thinking, i.e. rational, things, and Descartes could not imagine how rationality could be mechanised. One way of viewing the development of artificial intelligence programs of the past three decades is precisely as the attempt to show how rationality can be mechanised; how the capacity to make rational inferences can be embodied in a purely mechanical device.

Traditional approaches to this problem have operated on the following assumptions. Firstly, rationality is to be assimilated to one or another form of logic: deductive, inductive, abductive, and so on. Secondly, embodying rationality in a mechanical system amounts to building into that system certain structures and certain rules which transform those structures. The idea is that the rules correspond to the basic rules of logical systems, and that the structures are composed in such a way that they are capable of being transformed in the ways required by the logical rules. A simple example of this approach is provided by undoubtedly the most common type of artificial system: production systems. The basis of a production system is the rule of inference known as *modus ponens*: if P then Q. If the antecedent P is satisfied, then a so called production fires, and this causes the device to do Q. Chess playing computers are a common type of production system. This type of approach to mechanising rationality is sometimes referred to as the *rules and representations* approach, or, more commonly, the *symbolic approach*.

The symbolic approach was, for many years, the dominant research paradigm in both artificial intelligence and cognitive psychology. In the former this led to the proliferation of so called *expert systems*: systems that were restricted in their scope to a narrowly circumscribed task, playing chess or performing mathematical calculations for example, but within this domain were often strikingly good. In the latter, this led to attempts to model human cognition in terms of rules and representations. Problems with the symbolic approach, however, eventually became apparent. Most seriously, perhaps, there are certain tasks that symbolic, rule-based, systems are very good at. But there are also tasks at which they are very bad. In the former category are tasks such as mathematical calculations and formal reasoning in general. Roughly speaking, symbolic systems are very good at tasks which can be reduced to a series of sequential, essentially number-crunching operations. In the latter category are those tasks which do not, prima facie, seem reducible to such operations. These include visual perception or recognition tasks, categorisation, retrieving information from memory, and solving problems where the information provided is partial or inconsistent.

What is disturbing about this is not so much that there are certain things symbolic, rule-based, systems do not do very well, but the fact that the tasks human beings are good at are precisely those at which symbolic systems are bad, and vice versa. Humans are not good, nowhere near as good as most computers, at sequential number-crunching operations. We make, compared to even a simple hand-held calculator, an inordinate number of mistakes at long division. But we far exceed even the most complex computer in our ability to recognise faces, or to navigate our way across the living room, negotiating unforeseen events in the process. And this suggests that humans and symbolic systems accomplish cognitive tasks in fundamentally different ways. Thus while it might be possible to design extraordinarily efficient expert systems, these systems will provide very poor models of human cognition.

The tasks that humans are good at have a distinguishing characteristic: they all seem to be tasks that are reducible to *pattern-mapping* operations. Pattern mapping, in this sense, is made up of four different types of process. *Pattern recognition* is the mapping of a given pattern onto a more general one. *Pattern completion* is the mapping of an incomplete pattern onto a completed version of the same pattern. *Pattern transformation* is the mapping of one pattern onto a different but related pattern. And *pattern association* is the arbitrary mapping of one pattern onto another, unrelated, one. The reason symbolic systems are bad at these sorts of tasks is that they are not pattern mapping devices.

There is, however, a certain class of artificial system that is precisely such a device. The type of system is known variously as a *connectionist* system, *a neural network* model, or a *parallel distributed processor* (PDP).[11] Connectionism is an approach to modelling cognitive processes that is, or purports to be, neurally realistic. That is, the inspiration for the design of connectionist models is provided by the brain, conceived of as a network of interacting neurons. Accordingly, a connectionist network is made up of a collection of *nodes* or *units*, joined together by various *connections*. When energy is supplied to a node it becomes excited, and, if its level of excitement reaches a certain point, the node is said to be *activated*. Because of its connections to other nodes, the activation of one node can spread, or be propagated, to other nodes, and thus, the activation of a single node can, in certain circumstances, result in a pattern of activation becoming distributed across the network as a whole.

The specific details of this process are not pertinent here. What is important, however, is that connectionist systems are pattern mapping devices *par excellence*. And, in virtue of this, connectionist networks are capable of accomplishing a variety of cognitive tasks, including visual perception/recognition tasks, categorisation, recalling information from memory, and finding adequate solutions to problems with multiple partial and inconsistent constraints. In fact, connectionist systems are good at precisely the sorts of tasks at which humans are good, and bad at precisely the sorts of tasks at which humans are bad. And this has suggested quite strongly that connectionist networks might provide more realistic models of human cognitive processes than traditional symbolic systems.

The story, so far, then, is that human beings and connectionist networks seem, broadly speaking, to be good at the tasks that can be reduced to pattern mapping operations. The tasks that humans and connectionist networks are comparatively bad at include, most notably, logical and mathematical calculations and formal reasoning in general. These latter tasks do not, at first glance, seem to be a species of pattern mapping operation. Therefore, standard connectionist networks are hard pushed to deal with them.

What is important for our purposes is the connectionist response to this problem. One prominent response has been to investigate the ways in which connectionist networks can be *embodied*. To embody a network in this sense is simply to give it access to information bearing structures external to the net, and allow it the capacity to manipulate and exploit these structures in relevant ways. In other words, many recent connectionist attempts to model formal reasoning processes depend essentially on an environmentalist model of cognition. Cognition, on this extended connectionist model, consists not just in various processes internal to the net, but also on processes of manipulating and exploiting external information bearing structures.

Consider, for example, Rumelhart, Smolensky, McClelland and Hinton's well known account of our ability to engage in mathematical reasoning.[12] In a fairly simple case of multiplication, say $2 \times 2 = 4$, most of us can just learn to see the answer. This, suggest Rumelhart et al., is evidence of a pattern completing mechanism of the sort that can easily be modelled by a connectionist network. But, for most of us, the answer to more complex multiplications will not be so easily discernible. For example, 343×822 is not easy to do in the head. Instead, we avail ourselves of an external formalism that reduces the larger task to an iterated series of smaller steps. Thus, we write the numbers down on paper and go through a series of simple pattern completing operations (2×3, 2×4, etc.), storing the intermediate results on paper according to a well-defined algorithm. Thus, for example, upon recognition of the pattern '$2 \times 3 = \ldots$', an embodied system is able to complete that pattern and then, crucially, write or record the numeral '6'. This external structure creates a new pattern for the system to recognise, and its completion and recording in turn direct the system to a further pattern to be recognised and completed, and so on.

In this way, a process which seems *prima facie* to require the building of mathematical rules and structures directly into the machine can be reduced to an *internal* process of pattern recognition and completion together with a process of manipulating *external* mathematical structures. The entire process can be regarded as a combination of internal pattern mapping operations and manipulation of external mathematical structures. And the role that was thought to be played by *internal* symbols and rules has largely been usurped by *external* symbols and rules. We can, of course, learn to do long multiplication in the head. But our ability to do this, Rumelhart et al. contend, depends on our ability to mentally model external structures and relations. The environmental component has logical and chronological priority.

Bechtel and Abrahamsen have extended this approach to logical reasoning.[13] Indeed, once the framework has been properly delineated, it is fairly clear that this general approach can be extended to *all* types of formal reasoning. The problem for connectionist networks is that formal reasoning processes – of which mathematical reasoning provides one example – do not seem *prima facie* reducible to pattern mapping operations. What seemed to be required was that, as in the case of a production system, we build into all systems capable of performing such operations formal rules of inference together with representations structured in such a way that they can be transformed according to such rules. However, if we grant the system *embodiment*, if we give it the capacity to manipulate and exploit structures in its environment, then a process of formal reasoning can in fact be broken down into a combination of *internal* pattern mapping operations and *external* manipulation of environmental structures. Given that human beings seem to be connectionist systems, or something very much like them, and given that human beings are embodied in the relevant sense, then it seems overwhelmingly likely that this is the way humans learn to master formal reasoning processes.

The attempt to mechanise rationality thus takes a new and surprising twist. Rationality, the conforming of one's reasoning processes to normative principles, is not something achieved exclusively in the head. It is also, partly but essentially, something that involves, either in its development or in its present execution, the creation, manipulation, exploitation and transformation of environmental structures. Rationality, contrary to Descartes and the tradition that comes from him, is not something that separates us from the world. It is, in a very important sense, *of* the world. Rationality is something achieved *in* the world.

Once again, as far as the subset of cognitive processes responsible for reasoning are concerned, there seems to be no theoretically relevant distinction between the employment and transformation of internal information bearing structures and the employment and transformation of environmental information bearing structures. An organism can process information relevant to a formal reasoning task by acting upon, and effecting transformations in, formal structures (or physical instantiations thereof) in its environment. And an insistence on focusing purely on the internal operations involved will yield a seriously distorted understanding of the nature of cognition.

8.5 The danger and that which saves (revisited)

At one time or another, it has been fashionable to regard the view of nature as simply a resource as an aberration of some sort. The nature of this aberration has varied from time to time. Whereas it was once fashionable to think of it as a result of a commodity fetishism produced by unfettered market forces, now the current orthodoxy has it that the instrumental view of nature is the product of a specifically masculine conceptualisation of the world. But if the arguments of this and the previous chapter are correct, there is a clear sense in which these sorts of views miss the point. To view nature as a resource may, indeed, be an integral feature of capitalism. And it may intrinsically be bound up with a masculine conception of the world. But these phenomena can, at most, be a superstructure erected on a much deeper foundation. The real roots of our domination of nature are neither capitalist nor masculinist, but lie in certain widespread features of biological development that are prior to the development of economic systems and conceptual systems, and common to both masculine and feminine organisms.[14]

An evolutionary strategy of adaptation that involves manipulation or exploitation of environmental structures can, typically, be adopted at less cost than non-manipulative alternative strategies. With respect to a given evolutionary task, then, the differential fitness of a creature which has adopted a manipulative strategy will, again typically, be greater than a creature who has adopted a non-manipulative alternative. Therefore, if the creatures are in competition for the same environmental niche, the creature with the manipulative strategy of adaptation will, typically, be able to outcompete the creature with the non-manipulative alternative. Creatures of the former type, therefore, should flourish at the expense of those of the latter. In this way, manipulation of the environment becomes incorporated into a creature's biological evolution. Should the creature then go on to develop cognitive capacities, this facet of its biological development will (almost) inevitably be reflected in these capacities. Thus, examination of specific cognitive capacities in humans reveals cognitive processes to be essentially bound up with the manipulation and exploitation of structures in the environment. But a conception of the environment is one of the products of cognitive capacities. And even with due deference to the distinction between the origin of our cognitive capacities and the contents of the judgements that result from the exercise of such capacities, one can scarcely expect that a conception of the environment that derives from our cognitive capacities will be completely independent of the sort of environmental manipulation that provides the basis for those capacities. Consequently, we are committed in virtue of our natural history – in virtue of the forces, processes, exigencies, and constraints that went into our development, both biological and cognitive – to understand the environment as a resource.

There is an important sense, then, in which our understanding of the natural environment as a resource is, as Heidegger would say, a *destiny*. The waters of the instrumental conception of nature run deep, much deeper than has been commonly suspected. This is the danger. We have evolved, biologically and cognitively, as manipulators and exploiters of the environment. Due to our evolutionary development, and the resulting nature of our cognitive capacities, there is a very real sense in which we are constrained to see the environment as a resource, as something to be manipulated, exploited and transformed. It is difficult to see how one's thinking about the environment could not be tainted by this conception, since the very possibility of thinking is bound up with the possibility of manipulating and exploiting the environment. Understanding the environment as a resource is part of our natural history, part of what we are. It is not, then, something that can be easily discarded, as one would discard an unwanted garment.

But look where following the logic of the danger has led us. It has led us precisely to an environmentalist model of cognition, a model whose central features look like this. Firstly, there exist structures that are external to cognising organisms. Secondly, these structures carry information that is relevant to the completion of cognitive tasks. Third, organisms can process this information by acting upon, and effecting transformations in, these external structures. Therefore, fourth, not all the information processing effected by an organism in the performance of a cognitive task occurs inside the skin of that organism. Just as information processing can be achieved by way of manipulation of internal information bearing structures (mental representations), so too can it be achieved by way of manipulation of external information bearing structures. Therefore, fifth, cognition is not a purely internal process. Rather, it consists in the manipulation and transformation of both internal structures and environmental structures.

The environmentalist model of cognition entails the negation of the Cartesian mind/world distinction. Cognitive processes, processes such as perceiving, remembering, and reasoning are not purely internal processes; they are also, equally fundamentally, processes which occur in the world, they are processes *of* the world. Thus, the model not only allows us to break down the mind/world distinction, it allows us to do so in a way that is fundamentally anti-humanistic. The environmentalist model of cognition effects a dissolution of the Cartesian dichotomy not by the tired neo-Kantian method of pulling the world into the mind, but by the converse: pulling the mind into the world. And this, I think, opens up the possibility of developing an adequate theory of environmental value, a theory of environmental value that is predicated not on one or another form of idealism, but on a genuine environmentalist dissolution of Cartesianism. The environmental character of cognition might, therefore, allow us to develop a conception of environmental value that is neither subjective nor objective in character, that is immune to the standard objections to subjectivist and objectivist models, and which is not based on a humanist-idealist denigration of the ontological, epistemological and axiological status of the world.

Thus, the logic of the danger leads us precisely to that which saves. The danger, when properly understood, directs us to a properly environmental framework for understanding the value of nature. The transition from an environmentalist model of cognition to an environmentalist model of value is the subject of the next chapter.

9
Towards a Post-Humanist Theory of Value

If the arguments of the preceding chapters are correct, then we have a way of dismantling the mind-world dichotomy by *pulling the mind into the world*. In contrast to the overworked humanist-idealist tradition, we now have a way of understanding the mind as environmentally constituted; as not just connected to the environment but composed of it. We have a way of understanding ourselves as genuine *beings-in-the-world*. And, thus, we have, potentially, an axiological framework that does not necessarily doom the environment to secondary and derivative status. But how, exactly, do we move from an environmentalist theory of cognition to a genuinely environmentalist theory of value? How do we move from a conception of ourselves as genuine beings-in-the-world to a post-humanist conception of value in the world? This chapter aims to trace, in a way that is, admittedly, impressionistic and suggestive rather than detailed and complete, the logical contours of this move.

9.1 Requirements of the post-humanist model

We can, at the outset, sketch, at least in rough form, some of the central features of the post-humanist theory of value. These features can be inferred, or extrapolated, from two sources. Firstly, there is the environmentalist model of the mind itself. The central features of the model of environmental value will presumably follow closely, or be derived from, the central features of our model of the mind. Secondly, there are certain desiderata that we know a radical or environmentalist model of value will have to satisfy. These desiderata we can derive from the failure of traditional subjectivist and objectivist approaches to value. Consider, first, what we can infer about our model of value from the environmentalist model of the mind.

The environmentalist model of the mind developed in the previous two chapters can, roughly speaking, be described as follows. There exist structures that are external to – outside the skin of – cognising organisms. These structures carry information that is relevant to the completion of cognitive tasks. Organisms can process this information by acting upon, and bringing about transformations or alterations in, these external structures. And, therefore, not all the information processing effected by an organism in the performance of a cognitive task occurs inside the skin of that organism. Just as information processing can be achieved by way of manipulation of internal information bearing structures (or what are commonly known as *mental representations*), so too can it be achieved by way of manipulation of external (i.e. environmental) information bearing structures. Cognitive processes are typically defined as ones which involve the manipulation of information bearing structures in the performance of cognitive tasks. Manipulation of environmental information bearing structures, thus, qualifies as a type of cognitive process. Therefore, the skin has no relevance to understanding cognition. Cognition is not just something we do in our head. Cognition is fundamentally something we do in the world.

The development of the post-humanist conception of environmental value proceeds by way of two strands, corresponding to the two central elements of the environmentalist model of the mind. The first element was the claim that information exists in environmental structures. And accordingly, the first strand of the argument for an environmentalist conception of value consists in development of the idea that this value consists in a certain sort of information that is embodied in the environment. The second essential element of the environmentalist model of the mind was the claim that cognition consists, in part, of the manipulation, exploitation and transformation of relevant information bearing environmental structures. Cognition, then, belongs neither to the subject nor to the object; it straddles both. Accordingly, the second strand in the post-humanist conception of value consists in an account of our relation to environmental value that renders it neither subjective nor objective. The first strand, then, delivers a conception of environmental value. The second explains why this conception is radical rather than traditional, post-humanist rather than humanist.

Consider, now, what can be inferred about the nature of a post-humanist model of value from the failure of traditional subjectivist and objectivist accounts. The fairly sophisticated subjectivist and objectivist theories of Callicott and Rolston respectively can, with I think considerable justification, be regarded as attempts to pull off what would be a very impressive conjuring trick. But to see the precise nature of the trick, and just how impressive it would be if successful, it is better to first focus on simple versions of subjectivism and objectivism and the reasons for their failure.

As we saw in Chapter 3, simple objectivist accounts of environmental value run directly into a problem of arbitrariness. Objectivism, at least in its naturalistic form,[1] puts forward various features of the environment as the basis of intrinsic value, but fails to adequately explain why these features are valuable. The root problem here is the divorce, inherent in objectivism, between *value* and *valuing*. In thinking of value as objectively present in the environment, objectivism establishes an unacceptable distance between the value of the environment and its being valued by valuing organisms. And, for this reason, the objectivist's claims about the basis of environmental value appear arbitrary. The response of more a sophisticated objectivist such as Rolston is to try and re-establish the connection between value and valuing. But, as we saw, Rolston's proposal struggles to take us any further than biocentrism: the claim that life is intrinsically valuable.

Simple subjectivist accounts face the converse problem. Subjectivist models begin with valuing, and try to construct an account of environmental value on this foundation. Such accounts run into what is, in all essentials, a problem of normative force. They can point out that we *do* value the environment, but not that we *should*. Callicott's more sophisticated account is, in effect, an ingenious attempt to restore a dimension of normativity. But, as we saw in Chapter 4, his account fails for several reasons.

The root problem with both accounts is the same. They are founded upon a separation of value and valuing, and then are hard pushed to restore the connection that is seen to be required. Thus we seem ultimately to be left with a choice between an objectivism that tells what is valuable about the environment but not why we should value it, and a subjectivism that tells us what we do value about the environment but not why we should. And the conjuring trick is, then, this: to bring value and valuing together. This would be quite a trick. But it is something the environmentalist model of the mind might allow us to perform.

9.2   Valuing the environment

The central desideratum of a post-humanist model of environmental value is, then, that it re-establish the essential connection between the value of the environment and our valuing of it. But it must not do this in a way that simply reduces one to the other. If we go down that road, we simply arrive back at a traditional account, whether subjectivist or objectivist. On the other hand, a satisfactory post-humanist model will almost certainly not be characterised by a wholesale rejection of traditional subjectivist and objectivist accounts, but, rather, by its *rehabilitation*; by a reincorporation and reinterpretation of their best features. A successful post-humanist model of environmental value, that is, should be gauged not on the extent to which it neglects tradition, but to the extent that it banishes the weaknesses of this tradition and incorporates its strengths in a new, radical, form.

The task is, with respect to the environment, to bring together value and valuing. It is the subjectivist tradition that has most to say on valuing, and this section is concerned with the extent to which subjectivist accounts can be rehabilitated in post-humanist form. The following section deals with the notion of value, and the rehabilitation of the objectivist tradition.

The salient feature of the subjectivist model that is to be rehabilitated in radical, post-humanist, form is Callicott's important distinction between, in effect, the *origin* and the *content* of an evaluation.[2] Callicott correctly points out that to make a claim about the origin of value is not, at least not directly, to say anything about which objects possess value. In particular, the claim that the origin of value lies in various subjective feelings or sentiments does not entail that the only things which possess value are those feelings or sentiments. A claim can have an origin in subjective feelings without being *about* or *directed towards* those feelings. To suppose that there is a straightforward inference from claims about the *origin* of environmental value to claims about its *content* is to commit one version of the *genetic fallacy*.

The particular version of the genetic fallacy relevant to the development of the conception of environmental value in this chapter is the fallacy of inferring *instrumental content* from *instrumental origin*. To say that the origin of an evaluation is instrumental does not entail that the content of that evaluation is instrumental. To see this, consider Callicott's example, outlined earlier, of the parent's evaluation of his/her child.[3] Some things, such as money and medicine, we value instrumentally. Parents, however, do not, typically, value their children in the way they value money or medicine. Parents are supposed to value their children, and in normal cases do so value them, in a way that is independent of any selfish goal of the parent. Although a child may indeed have instrumental value for a parent, its primary value is independent of its instrumental character. In other words, we can value things in two clearly distinct ways. We can value a thing instrumentally, or we can value it inherently or intrinsically. This is a difference in the *content* of our valuing.

It might be tempting, at this point, to reply to Callicott in the following way. The valuing of a child by a parent, while it may appear to be an example of non-instrumental valuing, is really merely a disguised form of such valuing. The reason is that this valuing of one's offspring is based on a natural bond of familial affection, and this affection is instrumental in character; it is instrumental not relative to the parent, but to the parent's genes. That is, familial affection has been selected for because of the role it plays in preserving the gene line of parents. Parents may not value their children in the way they value money or medicine, but their genes value them in precisely this way: as an instrumental end to their own propagation. But to object to Callicott in this way, would be to miss the point of his distinction. The point is that the claim about what genes do or do not value is a point about the *origin* of an act of valuing performed by the parent. One cannot move from this claim to a further one about the *content* of this act of valuing performed by the parent.[4]

In other words, we can allow that the valuing of a child by its parent has an instrumental origin deriving from its role in the propagation of the gene line, but deny that this instrumental origin thereby affects the content of the act of valuing. Therefore, we can allow that it is a genuine case of intrinsic valuing, and not merely a disguised form of instrumental valuing. In this case, at least, the valuing by a parent of its child is a form of valuing that has instrumental origins but whose content has outgrown these origins.

This is important because it is at least arguable that *all* valuing has an instrumental origin. Individual human beings, for example, do not typically value themselves only instrumentally. It would be a strange person indeed who valued himself only because of the contribution he made to society, or to the quality of the lives of the people around him. Most of us seem to value ourselves intrinsically. But a plausible case could be made out for the claim that the reason we value ourselves in this way, indeed the reason any creature values itself, ultimately derives from the role the survival of an organism plays in the propagation of its genes. That, after all, is why there are organisms in the first place. This does not entail, of course, that the only things that are of value are genes, or that the axiological yardstick by which actions and events are to be measured is the survival of our genetic line.

To suppose that it does entail this would be to confuse the origin of our valuing with the content of our valuing. It would be to commit the genetic fallacy.

If it is indeed plausible that all valuing that is intrinsic in content is instrumental in origin – and it is genuinely difficult to imagine where else valuing may come from – then we should not hold our valuing of the environment up to any higher standard. I shall try to show that our intrinsic valuing of the environment does indeed have an instrumental origin, one buried so far back in the mists of biological time that we have forgotten it, metaphorically speaking of course. But this is not in any way to denigrate this valuing. On the contrary, it merely puts it on a par with, arguably, any form of intrinsic valuing.

9.3 Value in the environment

The post-humanist account of environmental value also rehabilitates certain features of the objectivist account. In particular, the objectivist intuition that value is something genuinely *in* the environment, and not merely something projected on the environment through human acts of evaluation, or those of other sentient creatures, is an intuition that the post-humanist account is capable of safeguarding. This section explains how.

The first essential element of the environmentalist model of the mind lies in the concept of *information*. Information exists in the environment. It can also be contained in the head. But, fundamentally, information is a matter of nomological dependence; and the relevant dependencies can be instantiated both in the head and, crucially, in the world. The skin is of no relevance to the understanding of, or the location of, information. There are, that is, at least some structures that carry information independently of whatever is going on inside the head of cognising organisms. Some of these structures carry information independently only of the mental states of *some* cognising individuals. Language provides an obvious example here. A word can have the particular meaning it has even if I am ignorant of or mistaken about that meaning. Its meaning is, thus, independent of my psychological attitudes. Other structures, however, carry information independently of *all* cognising individuals. The information contained in what Gibson calls the *optic array* is there independently of whether there is a perceiving organism around to detect and appropriate it. Indeed, the information would be embodied there even in a world devoid of perceiving organisms. The type of structures relevant to the following discussion of environmental value are those which carry information independently of all cognising organisms.

The first central claim of the post-humanist conception of environmental value, then, is that this value is to be identified with a certain type of information. This information is embodied in certain environmental structures, and is so independently of the mental states and attitudes of any cognising organism.

In developing this account of environmental value, the first concept that needs to be introduced is that of an *affordance*.[5] Environmental value is, in a way which will be made clear, closely associated with what Gibson refers to as affordances of the environment. The affordances of the environment are, for a given creature, what it *offers* the creature, what it *furnishes* or *provides*, whether this benefits or harms the creature. A relatively flat, horizontal, rigid, and sufficiently extended surface, for example, affords locomotion for many animals. It is stand-on-able, permitting an upright posture for quadrupeds and bipeds. It is, therefore, also walk-on-able and run-on-able. A non-rigid surface, like the surface of a lake, however, does not afford support or easy locomotion for most medium-sized mammals. It is not stand-on-able, but sink-into-able. Different substances of the environment have different affordances for nutrition and manufacture. Different objects of the environment have different affordances for manipulation, exploitation, circumventing, etc. In addition to substances and inanimate objects, the environment is also made up of other animals, and these afford a rich and complex set of interactions – sexual, predatory, nurturing, fighting, playing, co-operating and communicating. It is in the notion of an affordance that the deep roots of value lie.

For our purposes, the following features of affordances are most relevant. Firstly, affordances are *relational* entities: they exist only relative to given organisms. Thus, the surface of a lake affords neither support nor easy locomotion to a horse, but it offers both of these to a water bug. Thus, to speak of an affordance is to speak elliptically; an affordance exists only in relation to particular organisms with particular needs and capacities.

Secondly, affordances are *not* subjective. The existence of an affordance does not in any way depend on the attitudes, beliefs, opinions or feelings of organisms. A cliff affords danger for humans, even if one is of the opinion that one can fly. A lake surface, for humans, is sink-into-able, not walk-on-able, even if you believe otherwise. Moreover, it is true that a cliff affords danger for humans even in a world where there are no humans. That is, a cliff is the sort of thing a human could fall off, and consequently damage him- or herself, if any human should encounter it. The fact there are no humans around to encounter the cliff in no way impugns the truth of this statement, since it is conditional in form.

Third, the concept of an affordance is an essentially *normative* or evaluative one. To represent a feature of the world as an affordance is a form of evaluation. Affordances exist because creatures, such as humans, have certain *needs* which the environment is, or might be capable of meeting, and certain *capacities* that can be employed towards the satisfaction of those needs. The truth of the claim that a cliff affords danger for human beings does not require the actual existence of human beings, nor therefore, the existence of human beliefs, attitudes, opinions and the like. However, what it does require is that human beings would, *if* they exist, possess certain needs and capacities, both positive and negative. Thus humans would have the need to avoid certain sorts of bodily damage that can be occasioned by sharp impacts, and they lack certain capacities, such as the ability to fly. Affordances of the environment exist, for humans, because of the constellation of needs and capacities that such humans bring with them to this environment. And the concept of an affordance is, therefore, an essentially evaluative one.

The value in the environment, I shall argue, is closely associated with the affordances of the environment. It is not, however, to be directly identified with such affordances. Rather the claim is, first, that the environment possesses certain *indicators* of environmental affordances. There are certain features of any environment that are systematically, or nomically, connected with the affordances of that environment. These indicator features, therefore, carry the information that the environment possesses certain affordances. And, second, the value in the environment consists in the information embodied in these features. The environment possesses value in virtue of carrying certain sorts of information, and it carries this information in virtue of nomic correlations between indicators and the affordances they indicate. The question, then, is what sort of things these environmental indicators might be?

A start at delineating the concept of an indicator of an affordance can be made by examining the significant amount of work done by psychologists investigating the factors that underlie environmental choice. Underlying this work is a basic evolutionary argument: natural selection should have favoured individuals who were disposed to settle in environments likely to afford the necessities of life, but avoid environments with poorer resources or posing higher risks.[6] Ancestors who were good at identifying the features which make an environment a good one in which to live would flourish at the expense of those who were not. The needs of our ancestors were essentially the same as our current needs – to find adequate food and water, and protect themselves from the physical environment, from predators, and from hostile conspecifics. Our present mechanised, urban, environment furnishes these needs in rather different ways than the environment in which we evolved. Nonetheless, the number of generations who have lived in urban environments is vanishingly small compared to the number of generations spent as hunter gatherers; far too small to produce substantial evolutionary changes in our behavioural response patterns to physical features of the environment.

This evolutionary argument provides the basis of the well known *savannah hypothesis*.[7] The savannahs of tropical Africa, the presumed site of human origins, afford high resource potential for large, ground-based primates such as ourselves. In tropical forests, for example, nourishment is primarily afforded in and by the canopy, and a ground-based omnivore, in these environments, functions largely as a scavenger, gathering up bits of food that fall from the more productive canopy. In savannahs, however, trees are scattered and much of the productivity is found within two metres of the ground where it is directly accessible to a ground-based omnivorous ape. Biomass and production of protein is also much higher in savannahs than in forests. The savannahs also afford distant views, and the low ground cover favourable to a nomadic lifestyle. The savannah is, thus, an environment that provides what we need: nutritious food that is relatively easy to obtain, trees that afford protection from the sun and escape from predators, long unobstructed views, and frequent changes in elevation that allow us to orient in space. Evolution, it is assumed, should have equipped us with mechanisms that aid adaptive responses to the environment, and thus we should prefer savannah-like environments to other types of biome. This hypothesis has been experimentally confirmed. For example, in a series of experiments, a cross-cultural selection of individuals, when presented with black and white photographs of a variety of environments, consistently expressed a preference for savannah-type environments over others.

However, a preference for one particular type of environment over another will be grounded in certain more general or higher-order features possessed by one but not the other, or possessed to a greater degree by one rather than the other. These features would be indicators or indices of environmental affordances. They would indicate that the environment affords eating, drinking, shelter, protection from predators, and so on. And what is important for the purposes of a model of environmental value is not our preference for specific types of environment, but our predilection for the environmental indices which underlie this preference. What sorts of features would have this indicational role?

One example can be gleaned from an important set of experiments conducted by Orians and Heerwagen.[8] These experiments tested people's responses to the shapes of trees (specifically *Acacia tortilis*) found on savannahs. The significance of tree shape is that they are correlated with the quality of the savannah in question. In high quality savannahs – savannahs yielding in abundance the features listed above – the acacia tree, for example, has a spreading multi-layered canopy and a trunk that branches close to the ground. In wetter savannahs, the acacia has a canopy that is taller than it is broad with a high trunk. And in very dry savannah, the acacia becomes dense and shrubby looking. Orians and Heerwagen conducted a cross-cultural study of behavioural responses to trees with subjects from Seattle, Argentina and Australia. Subjects were asked to rate the attractiveness of each of the trees shown in photographs. The trees rated as most attractive by all three groups were those in which canopies are moderately dense and with trunks that bifurcate near the ground. In other words, all three groups of subjects showed a marked preference for trees that were indicative of high quality savannah. Trees with high trunks and either skimpy or very dense canopies were rated as least attractive by all three groups.

The shape of trees, then, provides one example of an indicator of environmental affordances. A certain tree-shape indicates that the environment is of higher quality than other similar environments and is, thus, more likely to afford adequate nutrition, shelter, and the like. This study, then, confirms the hypothesis that evolution has provided us with mechanisms that enable us to distinguish high quality environments from low quality ones, and that these mechanisms work by detecting certain higher-order features that are related in systematic ways to the affordances of the environment. Notice that the information picked up by the mechanisms is relational in at least two ways. Firstly, they are relative to a particular type of organism. What is a high quality environment for us is not for an arboreal monkey. Secondly, they are relational in the sense that they do not detect that an environment is good as such, but only that it is better than certain specified alternatives.

Tree shape is a pretty concrete indicator, and its indicational role is, accordingly, fairly specific. However, indicators can vary in their level of abstraction and, consequently, in the generality of their indicational role. That is, it is possible to discern a nested *hierarchy* of indicational environmental features; the higher-order the feature, the more general its indicational role. What might a higher-order indicator look like? Several examples are, in fact, provided by Orians and Heerwagen's more general theory of environmental choice. They distinguish three stages in environment choice.[9] Stage 1 choices are made on the basis of what we might call the *richness* of the environment. The richness of the environment in this sense is made up of such features as spatial configuration, depth cues, and certain classes of content, such as water or trees. These features are all indicative of the ability of the environment to meet human needs. The spatial configuration of the environment consists in graded features such as the degree of openness of an environment (e.g. desert versus closed forest). The spatial configuration of an environment is indicative of various affordances, positive or negative. An open environment, for example, does not afford cover from predators or hostile conspecifics. But, on the other hand, a closed forest leaves one perhaps fatally susceptible to ambush by those same predators or conspecifics. Depth cues, such as would be provided by the presence of hills or other raised vantage points, afford rapid assessment of distances, which are of value in determining the time required to cross open spaces, and the distance to potential prey or places of protection. And specific contents, such as the presence of water and trees, are quite obviously fairly reliable indicators of the availability of basic affordances such as eatability, drinkability, etc.

Stage 2 environment choices are made on the basis of information gathering factors, and centre around what we might call the *coherence* of the environment. In stage 2, the individual explores the environment to learn more about its potential to provide resources, and to see whether the prognosis provided by stage 1 factors will be borne out. Being a far-ranging and yet home-based organism places considerable priority on way-finding. Exploration of one's environment in order to get to know it well enough to range widely and yet not get lost would, thus, be an important element in a larger survival strategy. Such exploration would not only provide knowledge about critical resources such as water or edible plant material, it also might lead to a faster and more appropriate response in an emergency. At the same time, the importance of this knowledge creates a potential conflict. While the benefits of the knowledge afforded by exploration of new routes within a territory is quite clear, such exploration also affords dangers that derive precisely from the unfamiliarity of these routes. From this perspective, then, an ideal environment is one that is complex or rich enough to be promising from the point of view of its ability to afford essential resources, but not so complex as to be unreadable. The best trade-off between richness and readability is achieved if the environment possesses coherence, or instantiates what Humphrey calls repeated or *rhyming* patterns.[10]

Stage 3 of environment selection concerns the decision to stay in the environment for a significant length of time. Here, the crucial environmental indices are what we might call the *integrity* and the *diversity* of the environment. A suitable environment must contain a mixture of terrains that afford opportunities for all of the activities that are required to be performed if the environment is to be occupied for a significant time. Thus, the environment must be suitably diverse. However, not all of the activities required to be performed can be accommodated by a single terrain. A good foraging area might be a poor resting area, and so on. For this reason, the terrains must be related to each other in suitable ways. If, for example, the individual terrains are separated by large distances, too much time and energy will be lost in transit between terrains. The integrity of the environment pertains to the relations – spatial and temporal – between terrains. A good environment is one that possesses sufficient integrity as well as diversity.

Richness, coherence, diversity and integrity, in the senses explained above, provide examples of higher-order environmental indicators. These are more abstract features of the environment than, for example, tree shape, and, consequently, less specific in their indicational role. Are there any examples of yet higher-order indicators? Yes, but for examples of them we must switch our focus from environmental choice theory to ecology.

Here is a simple (idealised) example of what ecologists refer to as *succession*.[11] Imagine a lake with aquatic plants growing around the waterline. The roots of these marginal plants will, slowly but often inevitably, trap silt and soil, and this will build up over time so that the edges of the lake gradually move inwards. The aquatic plants will, therefore, move with the edges of the lake, and this leaves room for new species of plant to colonise the drier land effectively reclaimed by their aquatic cousins. These new plants will include marsh-loving trees such as hazel and alder, and these have larger water requirements so that more water is drawn from the soil. Moreover, with these trees come larger root systems, and these hold the soil more firmly together. Thus, the trees modify the ground in such a way that it now becomes suitable for colonisation by yet further species of tree, trees which like to keep their feet considerably drier – trees such as birch and ash. This sort of process is known as succession (more precisely, it is known as *obligatory* succession to distinguish it from the more transient but periodic disruption brought about by fire, disease, etc.). If this process of succession continues, the lake will eventually disappear, being replaced with woodland. And the pioneer species, the aquatic plants inhabiting the margins of the lake, will, of course, disappear in the process.

The transformation of the environment brought about by pioneer species, such as our aquatic plants, thus, seems positively suicidal. As F. E. Clements observed, back at the beginning of the century, 'Each stage reacts upon the habitat in such a way as to produce physical conditions more or less unfavourable to its permanence, but advantageous to the invaders of the next stage.'[12] This does not mean, of course, that there is some grand design underlying ecosystem development. The aquatic plants do not in any way sacrifice themselves for the greater good of the ecosystem. Rather, each aquatic plant is simply to maximise its own chances of survival. It just so happens that in doing so they ensure their own destruction.[13] Of course, it would be much better for the aquatic plants if they could somehow find a way of regulating their own population density to the extent that the lake is preserved. But, as we saw in our earlier imagined case of the hawks and the doves, such regulation would not be an evolutionarily stable strategy: it would be fatally susceptible to treachery from within.

The above account of a smooth successional process resulting in a stable and enduring state of *maturity* has taken something of a beating in recent years, but for reasons that are largely tangential to our concerns. For what is important for our purposes is the picture that underlies this account, and not the specific empirical details of the account itself. And the picture underwrites ecological theory in general, and not just this particular application of it. The picture is one of a network of indices; of features that carry information about the condition of the ecosystem. Thus, for example, the standard account of succession identifies interesting differences between mature ecosystems and developing ones, including the following:

1 The total amount of organic matter present in a mature ecosystem is larger than in a developing one.
2 Species diversity is greater in a mature system than in a developing one.

3   Biochemical diversity is higher in a mature ecosystem than in a developing one.
4   Stratification and spatial heterogeneity are far more organised in a mature system than in a developing one. That is, there is greater diversity of patterns in a mature ecosystem.
5   Food chains tend to be complex and web-like in mature ecosystems, whereas in developing ones they tend to be linear.
6   Stability (i.e. resistance to external perturbations) is high in a mature system, low in developing systems.
7   Entropy is low in mature systems, high in developing ones. That is, mature systems are more structured and complex than developing ones.[14]

The details of the specific differences between mature and developing ecosystems are not of concern here. Rather, what is of interest is the picture that underlies the successional model. The picture is of a web of nomically related features, all of which carry information about the character (i.e. developmental state) of the environment. The amount of organic matter present, species diversity (both in terms of variety and equitability), biochemical diversity, stratification and spatial heterogeneity, complexity (both of gross physical structure and energy distribution) and stability: all these are nomically connected with the developmental condition of the environment, hence all carry *information* about that condition. These features are relatively abstract, hence higher-order, indices of the state of the environment. They are, thus, at least indirectly, indices of what the environment is likely to *afford* human and other animals.

Now, the basis of environmental choice theory is that a creature which is able to detect what the environment affords it is going to do better in the survival game than a creature that is not. That is, it seems plausible to suppose that evolution will have built into certain creatures, specifically those creatures that are capable of deciding where to locate themselves, mechanisms which allow them to discriminate good environments from bad ones. And such mechanisms will, at some level, have to work by way of a sensitivity to the features of the environment – perhaps features such as those listed above – that are reliable indicators of what the environment affords or is likely to afford. Evolution equips relevant creatures with a sensitivity, an ability to detect, some portion of the web of nomically related features that carry information about its environment. And, if this is so, then our judgements of the value of the environment will be judgements predicated on the presence or absence of those features that are reliable indicators of the affordances of the environment.

This is not to say that any given creature will be able to detect the web in its entirety. Such sensitivity would, from an evolutionary point of view, be overkill, and would be bought only at a cost that would outweigh its benefits. Rather, evolution will equip a creature with sensitivity to some portion of the web, a portion of the web that, perhaps, carries information particularly relevant to the well being of that creature. Thus, the features that the environmental choice theorist tries to identify as guiding our evaluation of an environment must, ultimately, constitute some portion of the web of nomically related features that carry information about the character of the environment. That part of the web to which evolution has given us humans access is that aspect of the environment that indicates its value or worth to us. And it is these features that we will regard as giving the environment value.

There is nothing in this account, of course, that implies that our evaluation of the environment must be a matter of conscious judgement or reflection. Our detection of the relevant features of the environment, for the most part, seems to be carried out by non-conscious cognitive mechanisms. This would be so for a variety of reasons, most notably perhaps that evaluation of an environment is such a basic and vital ability that it would be possessed by creatures not capable of conscious thought or reflection. But the non-conscious character of these mechanisms in no way counts against them. We know, for example, that the vast majority of processes involved in any cognitive operation are non-conscious or *sub-doxastic* ones. And there is no reason to require any more of the mechanisms and processes involved in environmental choice.

9.4   A post-humanist theory of value

There is value in the environment. This value consists in certain sorts of information, information that exists in the relation between the affordances of the environment and their indices. This information exists in the environment independently of the beliefs, atttitudes, opinions, or feelings of sentient creatures. It exists independently of the evaluative acts of such creatures. Indeed, it exists independently of its being detected or picked up by such creatures. This information exists even if there are no creatures to detect or appropriate it. The information is *there*. It is in the world.

What makes this information value, however, is the fact that it is valued by valuing creatures, or that it would be valued by valuing creatures if there were any around. And the reason creatures have come to value this information is because it essentially bears on, in fact is essentially composed of, affordances of the environment. The information is essentially bound up with what the environment offers, furnishes, or provides valuing creatures. The valuing of the environment, then, is an act which transforms information into value, or, more accurately, allows us to see information *as* value. Valuing the environment is seeing information as value. And our valuing of the environment is something that has an instrumental origin, but one that is sufficiently shrouded in the mists of our biological history that it has outgrown this origin and now possesses intrinsic content. In this respect, it is precisely like the organism's valuing of itself, or the parent's valuing of its child.

Moreover, once the content of our evaluation has sufficiently out-grown its instrumental genesis, the content becomes, as we might say, *iterable*. Just as we intrinsically value not just our child but anyone's child, even though our valuing has an instrumental origin in the pre-servation of our genes, so also, to the extent that we admire certain environmental features, originally because of their role in indicat-ing environmental affordances, we can go on to admire those features, or what we take to be analogues of those features, even when they obtain in environments that, for one reason or another, are not suitable for human habitation. We can admire the mosquito-infested swamp because of its own kind of diversity, integrity, stability and so on even if we realise that these features, in the particular context of the swamp, do not indicate a suitable place for habitation.

This, then, is an outline of what a post-humanist theory of environ-mental value should look like. Value exists in the environment as information and this information is seen as value because of the evalu-ative acts of organisms. Seeing information as value has an instru-mental origin, but its content has outgrown this origin; it is now intrinsic.

This post-humanist conception of environmental value is distinct from traditional subjectivist and objectivist conceptions, and avoids the problems that plagued these accounts. Consider, first, the difference between this and traditional objectivist accounts.

Given the need, stressed earlier, to rehabilitate rather than simply reject, certain key features of objectivism, there are certain similarities between the post-humanist account described above and traditional objectivist accounts. Indeed, a superficial perusal might lead one to assimilate the two accounts. Such an assimilation would be based upon the thought that, according to the post-humanist account, environmental value is constituted by certain features that are indicative of the affordances of the environment. If the relation between affordances and their indices consists in a certain type of information embodied in the environment, and if information is conceived of as being objectively present in the environment, in the form of nomological dependencies that exist independently of any cognising subject, then we seem forced to embrace the view that environmental value is an objective feature of the world.

This assimilation, however, would be superficial. The key difference between the two accounts can be explained as follows. Traditional objectivist accounts are based on the idea that the relation between the value exists in the environment and the valuing activities of organisms is an *extrinsic* one. Value exists as value whether or not it is ever recognised as such, consequently whether or not it is ever valued. On the present account, however, value exists in the environment in the form of information whether or not it is ever recognised as such, but it is only seen *as* value when it is valued. Valuing is what allows us to detect information *as* value. This may seem a minor difference, but, in fact, it is one significant enough to allow us to bypass the difficulties of objectivist accounts.

Traditional objectivist models, as we have seen, are plagued by a charge of arbitrariness, a charge that received typical expression in the form of a sceptical 'Why?' question. Why is richness, for example, intrinsically valuable? Why diversity? Indeed, as we saw earlier, the same sceptical question can be raised against more orthodox naturalistic forms of objectivism. Why, for example, is rationality intrinsically valuable? Why sentience? The problem, ultimately, is that it is very difficult to make sense of the idea that something can have value independently of its being valued, and this difficulty manifests itself in the appearance of arbitrariness. This is why, in one respect at least, orthodox accounts are at an advantage over their environmental counterparts. Rational things, and sentient ones also, typically value themselves, and this makes it easier to understand both how they can have value and how it might be this rationality or sentience that gives them this value. As we saw, Rolston's account of intrinsic value was based on this strategy of deriving value from valuing. It is just that Rolston's account was based on the idea that all living things, not just rational or sentient ones, value themselves.

The post-humanist account of environmental value developed here also works by forging a connection between the *value* of the environment and our *valuing* of that environment. On this account, valuing does not constitute value, since value exists objectively in the environment in the form of information. But valuing does allow us to see information *as* value. This approach gives us an explanation of why features such as richness and diversity, integrity and stability should matter to us: they matter precisely because they are very general indicators of important environmental qualities, qualities that most definitely do matter to us. In this way, the present account of environmental value is not vulnerable to sceptical 'Why?' questions of the sort that troubled traditional objectivist accounts. The charge of arbitrariness is answered because the gap between value and valuing, a gap inherent in traditional objectivist accounts, is explicitly disavowed by the post-humanist conception of environmental value.

Consider, now, the difference between the post-humanist and traditional subjectivist accounts of environmental value. Again, as we should expect, a superficial perusal might suggest that the post-humanist account collapses into a form of subjectivism. For it could be argued that the affordances of the environment are not objective features but, rather, entities essentially constructed and constituted by certain processes occurring in the cognising subject. Affordances, it might be thought, cannot be regarded as perceptually basic features of the environment. We do not literally see the affordances offered by a stone in the same way as we see its shape. We do not feel affordances of a stick in the way we feel its weight. And we do not smell the danger in the air in the way that we smell the smoke. While it might be appropriate to regard certain things as perceptually basic in that we see them directly, without any process of inference being involved, it is not plausible to include affordances as among these things. On the contrary, it seems that affordances are perceived not directly, but only by way of an intervening process of inference. We smell the smoke in the air, we infer that smoke indicates fire, we combine this with the belief that fire can be harmful to us, and we then infer the presence of danger. Perception, it might be argued, begins with the detection of gross perceptual features – shape, weight, odour, and so on – and it is only by way of an often complex process of inference that we arrive at the presence of an affordance. And this, it seems, leads us straight back to a *projectivist* model of affordances, hence a projectivist model of value. For, affordances now seem to be something added to the world after all. Affordances exist only because of processes of inference existing in human subjects, and it is these processes which, in effect, add affordances to the world. Affordances are, therefore, subjectively constituted items.

This objection, however, gains whatever force it has only from the Cartesian, internalist, way of thinking about affordances, and, indeed, about value in general. Either, it is assumed, affordances must be something whose nature is objectively present in the world. Or, they must be something whose nature is, at least in part, dependent upon processes occurring inside the heads or skins of valuing creatures. And this way of thinking about value pretty clearly requires the claim that it is possible to separate off, in a way that is relevant to the understanding of affordances, the entities existing in the world from the processes occurring in the head. And it is precisely this assumption that the environmentalist model of the mind undermines.

If the environmentalist model of the mind is correct, then the very processes of inference that supposedly add affordances to the world are themselves processes that, at least in many cases, will be made up of environmental constituents. They will involve, and often do so quite essentially, the manipulation, exploitation and transformation of information bearing structures in the environment. Processes of inference are themselves *worldly*. There is no basis for viewing them as essentially internal items. Hence there is no basis for viewing the role of inference in the construction of affordances as being an essentially internal operation the results of which are projected on the world. Hence, even if we allow that affordances are partly constructed by processes of inference, there is no need for us to fall into the Cartesian trap of assuming that this makes affordances subjective, internal, entities projected onto an outer screen.

This, ultimately, is why the affordance based model of environmental value developed here is not subjectivist. What rescues it from subjectivism is the environmental model of the mind. Cognitive or information processing operations of the sort thought to be responsible for the construction of affordances are not necessarily processes internal to valuing subjects. They are equally likely to be environmentally located and environmentally constituted: made up of actions by valuing organisms on information bearing structures that are instantiated in the environment. Hence, the whole idea that the perceiving subject makes a contribution to perception that is essentially internal in character but is then somehow projected onto the world must, if the environmentalist model of cognition is correct, be rejected.

Therefore, the post-humanist account of environmental value is not a subjectivist one. The value of the environment consists in information that exists genuinely and objectively in the environment. This information is not in any way subjectively constituted, and any temptation to suppose that it is rests, in the final analysis, on an illicitly Cartesian construal of the nature of the mind and its operations.

Consider, now, the advantages of this model over traditional subjectivist accounts. According to the post-humanist account, environmental value consists in a certain sort of information embodied in the environment, information which our valuing activities can allow us to see *as* value. This information does not depend for its existence on the subjective feelings, preferences, beliefs, opinions or sentiments of valuers. Therefore, failure to grasp the value of the environment is not, as it is on a simple subjectivist model, *simply* a matter of failing to possess certain sentiments.[15] It is, fundamentally, a matter of failing to pick up or detect certain information that exists in the environment independently of such sentiments. Thus, if someone fails to appreciate the value of the environment, they are, on this model, lacking in a way in which someone who simply fails to possess a sentiment is not. In particular, they are *ignorant* of a genuine feature of the environment: they have failed to pick up or detect certain information that is embodied in the environment. Ultimately, then, the advantage of this model over traditional subjectivist ones can be expressed as follows. On the model developed here, our relation to environmental value is, at least in part, a *cognitive* one, whereas on simple subjectivist accounts our relation to such value is a purely *affective* one. This means that our relation to environmental value on the present account is, whereas on the simple subjectivist account is not, subject to *normative* considerations. According to the model developed here, failure to grasp the value of the environment is precisely that: a *failure*. And this is something that it cannot be on a simple subjectivist model.

9.5   The environmental turn?

To a considerable extent, the account of environmental value developed in this chapter is simply a sketch, an attempt to outline the most general contours of a theory, rather than an attempt to provide a theory as such. In part, of course, this is because I only have a very vague idea of what a post-humanist model of value would look like. But also, in part, it is because much of the work to do in defending the post-humanist model consists in undermining the influence traditional models have on our thinking about the environment and its value. Traditional models have, to use a Wittgensteinian expression, much of the status of a *mythology*.

That is, they exert a vice-like grip on our thinking and theorising about the nature of environmental value. They organise our attempts at theorising by making us think: this is how things must be. In this case, much of the philosophical work that has to be done consists in making us realise that things don't have to be this way at all. We do not, in fact, have to think about the value of the environment in the way the traditional approach implies that we do. This, perhaps, is the primary function of the environmentalist model of the mind and the model of value predicated upon it.

In essence, then, as they are developed in the preceding chapters, the environmentalist model of cognition and its axiological counterpart constitute more an approach to thinking about the value of the environment than a theory of in what the value of the environment consists. And it is, I think, as an approach, rather than a theory, that the true significance of environmental thought lies. The problem of environmental value derives from a widespread idea heavily implicated in the modernist world view. The idea is that it is possible to separate off, in a theoretically meaningful way, the contribution that a cognising subject makes to cognition from the contribution the world makes to cognition. The world, it is generally thought, provides only the raw materials for cognition in the form of the causal impingements on the cognising subject necessary for sensation. Cognition, itself, is a purely internal process of a subject. Once we embrace this picture, we are forced to regard certain features of the world as essentially constituted by human cognitive activity. And this is the orthodox Kantian view. And it is precisely in the challenging and undermining of this Kantian view that the real philosophical (as opposed to practical) importance of environmentalist thought lies.

According to neo-Kantian humanism, to idealism in the broadest sense of the word, the world of our everyday experience, the world we love, admire, loathe or fear, is a construction of the human mind. The nature of the phenomenal world depends essentially on the nature of the cognitive structuring activities embodied in the human mind. The question of the status of value is just one instance of this more general view. The value of the *environmental turn* in philosophy consists not in the refutation of this neo-Kantian view, but in providing a completely different, indeed contrary, framework for thinking about the world and its value. Instead of beginning with the human mind and investigating how this constitutes the world, we begin with the world and investigate how this constitutes the mind. The notion of constitution, here, is not causal. To view the constitution of the mind by the world as a form of causal transaction is already to buy into the Cartesian mind/world division. To say that the world constitutes the mind here is not to say that certain features of the world cause certain features of the mind. It is to say something much more fundamental. Certain features of the world are literally *constituents* of the mind.

This, of course, is not a refutation of neo-Kantian humanism. But it does turn it on its head. If the structuring activities of the human mind are themselves made up of environmental structures and processes, then the whole question of whether the world and its value is constituted by the structuring activities of the subject or is objectively existent independently of those activities is moot. It is in the difference in starting point that, I think, the true importance and legacy of environmental thought lies. Cognition itself is not a purely internal process; it essentially involves action upon environmental structures. It is essentially worldly. The world is not, as neo-Kantian humanists would have it, something constructed by our cognitive activities, the world is already in, already among, these activities, constituting them, in part, as what they are. Therefore, the value of the environment is neither something *out there* as opposed to in here, nor is it something *in here* that we mistakenly suppose to be out there. These oppositions lose their significance once we realise that we do not have to think about value in this way. If philosophy, at least for the past three hundred years or so, can be regarded as the working out of one or another form of neo-Kantian humanism, then the environmental turn would turn philosophy itself on its head.

10
Perspectives on the Environmental Crisis: Social Ecology, Deep Ecology and Ecofeminism

Perspectives on the roots of today's environmental crisis can, it is generally thought, be divided into three opposed camps. Identifying the central tenets of each camp, however, is no easy task. There is considerable variation in the claims made even within a single camp, and this sort of variation is exacerbated by clear differences in the philosophical acumen, scientific scholarship, and commitment to even simple principles of logic exhibited by different members of a given camp. Nonetheless, roughly, very roughly, we can divide the camps along the following lines, lines that will be severely qualified as the discussion proceeds. The first camp regards the environmental crisis as rooted, in a straightforward manner, in the human domination of nature, and this is a primary form of domination from which other forms can derive. Thus, oppression of human by human is often, on this view, considered a secondary or derivative phenomenon; derivative upon the oppression of nature by humans. The answer to both forms of oppression, and whatever crises they spawn, is to *rethink* our relationship to nature, to reconstruct this relationship so that it is no longer an essentially aggressive, hence oppressive, one. The recommended reconstructive strategy involves some form of *identification* with nature: the idea of the isolated Cartesian self should be rejected and in its place substituted an *expanded* self, a self that is properly part of nature and not opposed to it. This sort of view is often referred to as *deep ecology*. Accordingly, I shall talk of the deep ecology camp.

The second camp regards the human domination of nature as a secondary phenomenon, derivative upon a primary domination of human by human. The roots of the environmental crisis, on this view, lie in human social hierarchy and market-driven society. And this crisis is to be resolved by way of a critique of intra-human hierarchy: a decentralised, non-hierarchical, society, it is claimed, would necessarily be inherently ecologically sound. This sort of view is often known as *social ecology*, after its best known exponent Murray Bookchin. Accordingly, I shall talk of the social ecology camp.

The third camp is perhaps the most conceptually variegated. It is, therefore, the most difficult to classify. It is perhaps best, at this point, to focus only on the less sophisticated expressions of this camp, more worthy expressions being introduced later on as discussion develops. Thus, my characterisation of the camp might, at this point, be regarded by some as a broad caricature of it. This may be true, but if it is a caricature, then, as I shall argue later, it is a caricature that has, in fact, been adopted by a considerable and influential portion of that camp. In any event, the acknowledged inadequacy of scope of this characterisation will be rectified later. The third camp, at least on this simple construal, agrees with the social ecology camp that the human domination of nature is a secondary phenomenon, derivative upon the domination of human by human. But, while there are various forms of human–human domination, one of these is primary; the domination of women by men. Thus, the domination of nature ultimately derives from the domination of women by men. Or, as it is often claimed, *androcentrism* is prior to *anthropocentrism*. Following common usage, I shall refer to this as the *ecofeminism* camp.

10.1 Social ecology

It is the social ecology camp that admits of least conceptual diversity, and it is, therefore, here that we shall begin. The relatively monolithic structure of the social ecology camp is due to the fact that it is dominated by its most influential exponent, Murray Bookchin.[1]

Social ecology is a political movement that draws on the radical political tradition of critique and seeks an analysis of environmental problems in terms of structures of human social hierarchy. The fundamental premise of social ecology is that the domination of nature by humans derives from the domination of humans by humans. The notion of one domination *following from* could mean many different things, but Bookchin seems to have at least the following things in mind. Firstly, there is a claim of *historical* priority. The domination of humans by humans was historically antecedent to the domination of nature by humans:

All our notions of dominating nature stem from the very real dom-ination of human by human...As a historical statement [this] declares in no uncertain terms that the domination of human by human preceded the notion of dominating nature.[2]

This we might call the *historical priority thesis*. Secondly, there is what we might call the *causal dependence thesis*. Not only is the domination of humans by humans historically antecedent to the domination of nature by humans, the latter mode of domination depends causally on the former. That is, humans came to dominate nature because – in a causal sense – humans first dominated each other. The relation between the two forms of domination is not just one of historical succession but, more importantly, causal dependence. Third, the reference to the *notion* of dominating nature, as opposed to simply dominating nature, also suggests a distinct, but related, thesis of conceptual priority. The idea is that our possession of the idea that we are dominating nature depends on a prior possession of the idea that we are dominating humans. This claim itself, however, can be divided into two distinct claims, one relatively innocuous, the other extremely strong. The innocuous inter-pretation of the claim is simply that we could not form the idea that what we are doing to nature constitutes domination unless we had acquired the concept of domination from our relations to human beings. This is a claim about concept *acquisition*; how we come to acquire the concept of domination. It claims that we acquired this concept first in the context of human–human interaction and later extended it to our relations with the environment. This is a relatively harmless claim, and may even be true. We can call it the *weak conceptual priority thesis*. This weak claim is, however, to be distinguished from a much stronger claim, a claim that Bookchin also appears to endorse. The concept of domination applies, in any real or genuine sense, only to human–human interactions, and its application to other domains is strictly derivative and metaphorical. This is a claim about the *content* of the concept of domination rather than its acquisition. The concept of domination is one whose content essentially relates to human–human interactions; its extension to human–environment relations is derivative upon this. We might call this the *strong conceptual priority thesis*. Finally, there is what we might call the *strategic priority thesis*. Bookchin claims that human liberation is strategically prior to, and must come before, the liberation of nature.[3] This is because the domina-tion of nature by humans is dependent, in the above senses, on the domination of humans by humans.

Because the domination of nature by humans is derivative, historic-ally, causally and conceptually, on the domination of humans by humans, and since the liberation of human oppression is strategically prior to the liberation of nature, environmental problems must be approached and resolved, Bookchin argues, through a critique of struc-tures of human social hierarchy: 'Ecology alone, firmly rooted in *social* criticism and a vision of *social* reconstruction, can provide us with the means for remaking society in a way that will benefit nature *and* human-ity.'[4] It is reasonably clear, then, that the coherence of Bookchin's posi-tion depends on the assumption that human domination of human is prior to human domination of nature. And, of the three priority theses outlined above, the historical priority thesis is clearly the most central. My focus, therefore, will be on this. What reason is there for believing the historical priority thesis?

In fact, not only is there very little reason for believing the historical priority thesis, there are overwhelming reasons for *not* believing it. One implication of the thesis is that nature was a non-exploitative domain before humans began to dominate first each other and, subsequently, the world around them. Manipulation and exploitation of the environ-ment are not basic features of organic life, but are something humans essentially invented by way of their manipulation and exploitation of each other. This is a common romantic myth, one which the history of philosophy throws out at us every now and then. But that is all it is: a myth. The problem with this myth is that it flies in the face of every-thing we know about evolution. The very theoretical foundations of evolutionary theory entail that this claim simply *must* be wrong.

Consider, for example, the central features of the evolutionary account of the origin of life.[5] The essential story is well known, so I shall be brief. Molecules of compounds such as water, carbon dioxide, methane and ammonia, under the influence of ultraviolet radiation, combine to form more complex molecules, in particular amino acids, the building blocks of proteins. This sort of process gives rise to the 'primeval soup' which chemists and biologists believe constituted the seas of three to four thousand million years ago. The organic molecules, again under the further influence of ultraviolet radiation, combine to form yet larger molecules. At some point in this process, a particularly remarkable molecule was formed by accident: a *replicator*, a molecule with the ability to create copies of itself. This is not as remarkable an accident as it might at first appear, at least not when we think on geological time scales. Imagine the replicator as a complex chain of building block molecules, a chain made up of smaller building blocks, and these smaller building blocks will themselves be abundant in the soup surrounding the replicator. Now suppose that each element in the chain has an affinity for its own kind of building block. If this were the case, then whenever a building block from the soup becomes adjacent to the element of the chain, it will tend to stick there. The building blocks which attach themselves in this way will automatically be arranged in a way that copies the original replicator. They would thus form a chain that is an exact copy of the replicator. The process could continue as a progressive stacking up of layer upon layer (as in, for example, crystals). Or, it could happen that the two chains split apart. In this case, there will now be two replicators, each of which can go on to make further copies. Replication is the first essential element of evolution.

If such a process were to continue, we would arrive at a population of identical replicas. However, as in any copying process, mistakes can happen. And, as miscopyings were made, the soup became populated not with identical replicas, but with several varieties of replicating molecules. Variation is the second essential element of evolution.

It seems that some varieties of replicator would spread more rapidly than others. This is for two reasons. Firstly, some varieties would be more stable than others, and would therefore tend to become relatively more numerous than those more likely to break up quickly. Secondly, and more importantly, some varieties of replicator were more *fecund* than others. Some replicators, that is, would be better than others at attracting building blocks from the primeval soup. Therefore, they would, all things being equal, make copies of themselves more quickly than other varieties. The primeval soup, containing a finite number of building block molecules, was not capable of supporting an infinite number of replicator molecules. Therefore, replicators that were better at attracting building block molecules would not only become more numerous than other less efficient ones, they would, in fact, thrive at the expense of their less efficient competitors. Less favoured varieties must have become *less* numerous because of competition, and some lines must have gone extinct. Competition is the third essential element of evolution.

A process of evolution, as Dennett points out, is an algorithmic process.[6] Part of what this means is that an evolutionary process is *substrate neutral*. Evolution, that is, is not necessarily a biological phenomenon, but happens whenever the following conditions are met: *replication, variation, competition*. These three conditions are necessary and sufficient for evolution to take place.

Without competition, there is no evolution. This is a truism in evolutionary circles. And the arguments developed in Chapter 7 are essentially just a working out of the consequences of this truism. If an organism manipulates its environment in the accomplishing of a task, then it gets the environment to do some of the necessary work for it. Therefore, it will have more resources (i.e. energy) left to accomplish other tasks thrown at it by the querulous environment. Therefore, with respect to this task at least, it will be differentially fitter than competitors that do not manipulate the environment, or do not do so as efficiently. There is nothing controversial here. That competition is one of the factors that drives evolution is a truism. That getting someone or something else to do some of your work for you puts you at a competitive advantage over others who do not is another truism. Yet Bookchin, it seems, is committed to denying one or other of these truisms.

How could he make this mistake? How could Bookchin claim that manipulation of the environment not only originates with humans, but is actually derivative upon human manipulation of humans? How could he not see that manipulation of the environment is something that came on the scene long before humans ever did? The answer is that he is working with a concept of evolution that is, frankly, laughable.[7] He takes it as a 'presupposition' or an 'intuition' that evolution is driven by an 'inherent striving' towards greater complexity, freedom, and subjectivity. The support he offers for this claim is simply that it is an 'intuition' based on the undisputed claim that there has been an observable trend to more complexity. There has been an observable trend to more complexity, more freedom, more subjectivity. Therefore, there is an inherent striving towards more complexity, freedom and subjectivity. This sort of inference is, at best, pre-Enlightenment, and has been decisively refuted by Darwin and his neo-Darwinist successors. Indeed, if one had to summarise the principle of evolution in one sentence it would be in terms of the negation of this claim. As we rapidly approach the twenty-first century (indeed, as this book comes out we have already presumably reached it), Bookchin, it seems, needs to wake up and smell the nineteenth!

There are, of course, ways of defining the notion of domination that makes it possible to rescue Bookchin from absurdity. Such a way is suggested by Bookchin when he rejects even the appearance of hierarchy in primate societies. The alpha male, for example, does not, according to Bookchin, dominate other members of his group because he does not rule 'through institutional forms of violence as social elites do.'[8] So, we avoid the conclusion that domination is rife in the natural world by breaking the connection between manipulation and exploitation, on the one hand, and domination on the other. The natural world, on this suggestion, is riddled with manipulation and exploitation but not with domination, since this requires the employment of 'institutional forms of violence'. Domination, therefore, is an essentially political activity, hence one that could not have preceded the development of human societies.

This is, in effect, the philosophical equivalent of the *oldest trick in the book*. One can always avoid an unwanted conclusion by redefining the terms that led you to it (after, of course, shouting 'look behind you!' first). Alas, such redefinition almost always comes at a cost. In this case, the cost is twofold. Firstly, Bookchin's position then becomes truistic, indeed almost tautological. Of course human domination, in *this* sense, of nature is subsequent to human domination of humans, since domination requires the establishment of a 'social elite' that can wield institutional forms of violence. That is, domination has now been defined in such a way that human 'domination' of nature must necessarily be subsequent to human domination of humans. Secondly, and more importantly, Bookchin's redefinition of the concept of domination works only by breaking the conceptual link between manipulation/exploitation and domination. Manipulation and exploitation of the environment are to be clearly distinguished from domination of it. However, in divorcing manipulation from domination, one also breaks the link between domination and environmental problems. Once we make domination an essentially political activity, involving the use of institutional forms of violence, then we no longer have any reason for thinking that it is domination that is at the root of our environmental problems rather than manipulation or exploitation. Why, now, suppose that our environmental problems stem from domination, understood as an essentially human activity involving institutional forms of violence, rather than the more biologically widespread manipulation and exploitation of the environment that has been incorporated into the natural history of human beings? The price of narrowing the scope of domination in the way suggested by Bookchin is that it robs the claim that domination is at the root of our environmental problems of any plausibility.

10.2   Deep ecology

The deep ecology camp encompasses a spectrum of writings, from authors of very different intellectual traditions and very different intellectual credentials. The result, inevitably, is that deep ecologist writings tend to diverge wildly in their quality. And, in many ways, it would be more than harsh to lump the lucid exposition and argument of some deep ecologists with the *shambalic* declamations of others.[9] My focus, in this section, is on the relatively unsophisticated end of the deep ecology market, more sophisticated versions being discussed towards the end of the chapter. With this focus in mind, it is possible to identify certain common themes running through these diverse writings, although emphasised to differing extents, that effectively constitute their deep ecological character and justify grouping them together.

1 Human domination of nature is the primary form of domination from which other forms derive. Human domination of humans is secondary to, and derivative upon, human domination of nature.
2 The aggressive and manipulative mode of human interaction with nature which makes up this domination is unhealthy. We need to 'rethink' our relationship to nature.
3 Rethinking this relationship involves the acquisition of an 'ecological consciousness' which allows us to 'identify' ourselves with our environment.

According to deep ecology, our relationship with nature has been a 'bad' or 'unhealthy' one characterised by domination, disrespect, disharmony. Therefore, we need to rethink this relationship; to remove the offending elements from it. According to Devall and Sessions, the primary project of deep ecology is to 'articulate a comprehensive religious and philosophical world view', a 'new ecological philosophy for our time',[10] the basis of which is provided by some sort of 'ecological consciousness'.[11] And this ecological consciousness seems, in effect, to be defined in sole opposition to Cartesianism.[12] This opposition has at least three distinct strands. Firstly, Descartes regarded the physical world as essentially a vast *mechanism*; a collection of intricately related but essentially inert, dead, pieces of extended matter. Against this, deep ecology posits *organicism*; the physical world is essentially a complex organism. Descartes regarded the nature and behaviour of any item in the physical world as *reducible* to the nature and behaviour of its parts. Against this reductionism, deep ecology advocates *holism*; the nature and behaviour of anything is bound up with its relations to nature as a whole, and is not amenable to reductive explanation in the manner Descartes thought. Descartes, as we have seen, is also a *dualist*; minds are essentially different kinds of things than bodies. The former are non-physical, spiritual substances, the latter lumps of brute, unthinking, extended substance. Against this dualism, deep ecology posits *monism*, the unity of mind and nature, not in the sense that minds are reductively explicable as physical mechanisms, but in the sense that nature is also spiritual. The result is that nature is *respiritualised* and, hence, *resacralised*.

The environmental problems we face today are, according to deep ecology, the result of the artificial separation of humans from nature. The key to rejecting this separation, hence rethinking our relationship to nature, lies in the identification of self with nature. This identification will underwrite a more harmonious relationship with nature, and the resulting resacralisation will underwrite a more respectful relationship. What, however, does it mean to *identify* oneself with nature?

The notion of identification is, in fact, ambiguous. Following Val Plumwood, we can distinguish three different senses of identification employed by deep ecologists: *indistinguishability*, *expansion* and *transcendence*.[13] According to the indistinguishability construal of identification, there are no boundaries between self and nature. For example, Warwick Fox writes: 'We can make no firm ontological divide in the field of existence...there is no bifurcation in reality between the human and non-human realms...to the extent that we perceive boundaries, we fall short of deep ecological consciousness.'[14] Of course, this claim must be correct for the simple reason that it is trivial. Everything that exists is made up of atoms, which are made up of sub-atomic particles, and so on. There is no firm ontological distinction between one collection of atoms and another, for the simple reason that they are both collections of the same sorts of thing, and that members of one group can in principle be transferred to the other. Every breath we take is testimony to this truism. The question is, however, what follows from this? Well John Seed, for example, thinks that a lot follows from this. Specifically, he feels that widespread policies of environmental protection follow on the grounds of self-interest:

As the implications of evolution and ecology are internalised... there is an identification with all life...Alienation subsides...'I am protecting the rainforest' develops to 'I am part of the rainforest protecting myself. I am that part of the rainforest recently emerged into thinking.'[15]

However, as we saw in Chapter 5, the absence of a firm distinction between self and world is compatible with a variety of attitudes towards that world. Fear or loathing of an uncooperative and obdurate part of oneself that one would like excised is, for example, another possibility. We are, of course, in the world in the truistic sense described by Fox, but being in the world in this extremely general and diaphanous sense is compatible with a host of more specific ways of being there: nurturing or exploiting, stewarding or dominating. And one's attitudes to the world, and the moral claims one perceives the world as making on oneself, are determined by these more specific modes of being-in-the-world. This, ultimately, is why the lack of any 'firm ontological divide' between myself and my local landfill does not engender in me any desire to protect this landfill, even as part of the landfill recently emerged into thinking.

The second interpretation of the notion of identification to be found in deep ecological writings is what Plumwood refers to as *expansion*. Identification, in this sense, does not mean 'identity with' but something more like 'empathy'. One identifies with the environment, in this sense, in the same sort of way that one might identify with the suffering of another, or identify with another on the basis of shared suffering. As Aldo Leopold says, 'One of the penalties of an ecological education is that one lives alone in a world of wounds.'[16] The problem with this suggestion should, however, quickly be evident. One can, of course, identify with the environment in this sort of sense. Many do so, and many more profess to do so. But the question is: why *should* we identify with the environment in this way? Environmental difficulties would no doubt be ameliorated if all of us identified with the environment in this way. But this gives us only a prudential, or human centred, reason for doing so. What deep ecology requires, pretty clearly, is not just to advocate identifying with the environment in the sense of empathising with it, but also to defend this advocacy: to show *why* we should identify with the environment. But this defence would seem to rest on an account of the value of the environment, and any advocacy of empathy with the environment would appear to be secondary to this. Therefore, if deep ecology simply amounts to the claim that we should identify with the environment in the sense of empathising with it, deep ecology is essentially vacuous.

The third interpretation of identification we can, following Plumwood, label *transcendence*. Thus, Fox, one of the more sophisticated deep ecologists, claims we should strive for impartial identification with our own particular concerns, emotions and atttachments.[17] In effect, Fox is presenting an environmentally oriented version of the universalisation condition constitutive of traditional human-based ethics. Thus, for example, a human centred utilitarian ethic might claim that we are to maximise the number of satisfied preferences, irrespective of whose preferences are satisfied, and irrespective of when they are satisfied. Thus, preferences are treated impartially. Whether a preference is mine or somebody else's is morally irrelevant. Similarly, a contractarian position might argue that we should begin moral deliberation from a hypothetical position of ignorance, an *original position*, essentially to guarantee this kind of impartiality. What tends to be overlooked however, is that the scope of this sort of universalisation is not, itself, universal. On the contrary, the principle of universalisation or impartiality is restricted to the domain of so called *morally considerable* individuals, where to say that an individual is morally considerable is just to say that it is the sort of entity to which the principles of morality can, in whole or in part, apply. There are various ways of trying to widen the scope of this domain. As we saw in Chapter 2, one strategy commonly employed by defenders of animal rights is to argue that the domain should be widened to include some non-human animals because there are no morally relevant differences between humans and at least some such animals. But, it is difficult, to say the least, to see how this sort of moral extensionism could be applied to the case of the environment: ecosystems, endangered species, and the like. And most advocates of a non-human based environmental ethic would, in any event, reject this sort of extensionism. Indeed, what makes environmental ethics so difficult, and so interesting, is the fact that it seems to require a completely different moral basis than human centred ethics, or extensions thereof. Whether or not this ultimately proves to be correct, the upshot is that Fox's transcendental version of identification will work only if it can be argued that the environment and its components are morally considerable entities. And once again we are led back to a quest for the value of nature. So, on this third interpretation, as on the first and second, the deep ecological concept of identification leaves the hard work undone.

Truistic, vacuous, or seriously incomplete. The options are unenviable. Whichever interpretation of the deep ecological concept of identification is adopted, the result is immensely unsatisfactory. But identification with the environment is the conceptual cornerstone of the deep ecological position. The position is, therefore, I think, in serious trouble. Should we witness the demise of at least the cruder versions of deep ecology, this, I think, would not be a bad thing from the perspective of the environmental movement as a whole. Whatever the motivations of its proponents, there is something deeply disquieting about deep ecology, especially in its more simplistic versions. It purports to be scientifically based, on evolutionary and ecological theory, but its development in the hands of influential proponents seems to require a total misunderstanding of the nature of such theories. Arne Naess, for example, claims that in nature, cooperation, 'the ability to coexist and cooperate' is more important than competition, 'the ability to kill, exploit, and suppress.'[18] Anybody who can make this claim has simply misunderstood the theory of evolution. Allied to its ignorance, denial, neglect, or disregard of the central scientific concepts that supposedly inform it is its affiliation with mysticism, either American nature mysticism or of an Eastern variety. Murray Bookchin may be a little unfair when he writes, 'Deep ecology was spawned among well to do people who have been raised on a spiritual diet of Eastern cults mixed with Hollywood and Disneyland fantasies.'[19] Nevertheless, in its cruder forms, the result of deep ecology is often the sort of 'touchy, feely' caricature of environmental thought that has, I think, done inestimable harm to the environmental movement, and will continue do to so for as long as it is promulgated and tolerated.

10.3   Ecofeminism

It would be a mistake to regard the term 'ecofeminism' as denoting a single position. Just as with deep ecology, ecofeminist writings vary widely in terms of content, and even more widely in terms of quality. It would be harsh indeed to lump together the thoughtful and trenchant writings of a Val Plumwood, for example, with the ravings of certain other ecofeminists.[20] And my focus in this section will be purely on the more misguided end of the ecofeminism market. More sophisticated, hence more realistic, versions of ecofeminism will be discussed towards the end of the chapter.

This immediately raises a methodological problem. Men who have criticised ecofeminism in the past have often been responded to in an *ad hominem* manner. Criticism of ecofeminism by certain males, that is, has often attracted the response that this is simply manifestation of a desire to keep women barefoot and pregnant, or as a manifestation of a belief that women should be seen and not heard, or whatever. The variegated nature of ecofeminist writings, in fact, invites this sort of response. One aspect of ecofeminism is criticised, but this criticism does not cover other aspects. The person (i.e. man) who raises the criticism is then accused of being ill informed, and the logical explanation for him allowing himself to write such a criticism, knowing himself to be so negligently ill informed is an uncontrollable desire to put women in their place. This response to male critique has appeared in print on several occasions.[21] It is difficult to know what to say to this response, and perhaps the best thing to do is ignore it. But, to avoid this sort of unpleasantness, let me just reiterate that in this section I am dealing with one particular aspect of ecofeminism, and my criticism is not meant to cover all ecofeminist positions or everything that has gone by the name of ecofeminism. Also, let me point out that the problems with the relatively unsophisticated brand of ecofeminism discussed here are problems clearly recognised by some of the more thoughtful ecofeminists.

Several ecofeminists either imply, or simply dogmatically assert, that androcentrism, the domination of women by men, is somehow prior to anthropocentrism, the domination of nature by humans. Ynestra King, for example, writes:

> It is my contention that the systematic denigration of working-class people and people of colour, women and animals is connected to the basic dualism that lies at the root of Western civilisation. But the mind-set of hierarchy originates within human society. It has its material roots in the domination of human by human, particularly of women by men.[22]

The claim that hierarchical thinking has its *material* (as opposed to conceptual) roots (particularly) in the domination of women by men, presumably entails that androcentrism historically precedes other forms of domination, including the domination of nature by human beings. And, in an earlier essay, King actually claims that this thesis is characteristic of ecofeminism in general.[23] In a similar vein, Sharon Doubiago claims that 'ecologists have failed to grasp the fact that at the core of our suicidal mission is the psychological issue of gender, the oldest war, the war of the sexes.'[24] And Vandana Shiva also calls androcentrism the 'oldest of the oppressions'.[25]

The problems with this position are hopefully clear. Indeed, we have encountered them already in the discussion of Bookchin's social ecology.[26] Manipulation and exploitation of the environment are pervasive features of the biological world. Given the central principles of evolutionary theory, they *have to be* pervasive features of the biological world. Any organism that is able to manipulate or exploit its environment – where this environment includes other conspecifics – is able to get that environment to perform some of its work for it, some of the work that must be performed in order for it to survive. Such an organism would thereby be at a selective or evolutionary advantage over organisms that did not manipulate their environment. This clearly preceded the domination of human females by human males since it preceded the evolution of the human race. If we want to look for *roots* to our environmental problems, it is surely here that they will be found, and not in androcentrism.

Of course, as with the case of Bookchin, one could always redefine the notion of domination to make it true that androcentrism is prior to anthropocentrism. As we have seen, however, the price of this sort of redefinition will be to not only make the claim tautological but also to undermine the thesis that it is androcentrism that is at the root of environmental problems. If the form of domination characteristic of androcentrism has no constitutive connection with manipulation and/or exploitation of the environment, then there is no longer any reason for supposing that it is this form of domination that underlies our environmental crisis.

Is there anything crucial to ecofeminism that rides on the question of whether androcentrism is prior to anthropocentrism? Does ecofeminism make sense only on the condition that the former is prior to the latter? I think not. And more sophisticated and plausible versions of ecofeminism are already recognising the claim that androcentrism is prior to anthropocentrism for the dogma that it is.[27]

10.4   The web of oppressions

Social ecology, deep ecology, and (the above version of) ecofeminism present us with a false choice. Social ecology and ecofeminism (at least in the form outlined above) claim that human–human domination comes first, although they disagree on the specific nature of this domination, and that human domination of the environment is derivative upon and secondary to this human domination of human. Deep ecology, on the other hand, claims that human–environment domination comes first, and that human–human domination is subsequent to this prior form. What all three accounts fail to recognise is that the history of human domination, both of humans and the environment, stretches back long before even the existence of humans. The forerunners of humans were manipulating and exploiting the environment long before humans had come to be. So, if we want the roots of human domination of the environment we will not find them in human domination of human. And neither will we find the origin of human domination of human in our relation to the environment. The roots of both stretch back deep into our natural history: the roots of human domination, in all its forms, lie in a past that had not even come to contain humans.

Understanding the roots of human domination is important, but not in the straightforward way sometimes imagined. It is sometimes thought, or implied, that the various forms of human domination or oppression stack up one on top of the other, like a house of cards, so that if we remove the foundational form of domination all the others will collapse with it. This picture is, I think, far too simplistic. Far more likely is that, to use an analogy of Plumwood's, oppressions form a *web*.[28] A web can continue to function, and indeed repair itself, despite localised damage to its parts. Various oppressions in such a web would not depend for their efficacy or stability on their relation to some foundational component, nor, consequently, would they be understandable in terms of such a foundation. Just as a web functions by having strands that pull in opposite directions, so too might some oppressions be fundamentally or qualitatively different from each other, yielding different consequences and requiring different solutions.

Whether or not this analogy is ultimately apposite, it is unclear, to say the least, what *removing* the root of human domination would amount to. How does one 'remove' part of one's natural history? How does one cut away from oneself a historical narrative that began long before one was ever born? How does one cut away from human nature something that was responsible for the formation of that nature; indeed, something that is inevitably responsible for the formation of all biological natures? At most, we can *understand* how our nature is necessarily bound up with dispositions to manipulate and exploit the environment. In doing so, we understand the danger. The inevitable manipulative and exploitative tendencies possessed by any biological life form became embodied in a creature who possessed an opposable thumb, and who subsequently underwent a process of massive encephalisation; consequently a creature who became very, very good at expressing these tendencies. And understanding the danger, in this context, is all that metaphors of removal or cutting away can possibly amount to.

The various forms of domination and oppression embodied in human relations to the environment and among each other are, then, forms that are ultimately built upon this sort of biological substructure. But there is no reason to suppose that the various dominations and oppressions that derive from this source are in any way qualitatively reducible to it. To use an analogy made famous by Wittgenstein, the source of the plant is the seed, but the various features of the plant, leaves, stems, etc., need not correspond to discrete features of the seed.[29] Or, to return to Plumwood's metaphor, the features of the web, features in virtue of which it flourishes, do not correspond in any reductively straightforward way to the features of the spider that produces it. Dominations and oppressions that derive from a single source can develop in quite different ways, can take a variety of forms, can yield quite distinct consequences, and require very different solutions.

Once we appreciate this, and once we excise their misconceived pre-occupation with foundations, social ecology, deep ecology and eco-feminism can be understood as having a quite distinct function and value. Each, in its own way, yields insight into some restricted part of the web of domination. Since the strength and stability of any web consists in its overall structure, the web can survive localised damage to its parts. Destroying the web, therefore, requires a multi-dimensional attack. But what would rob such an attack of its efficacy and coherence would be if one prong in the attack misunderstood its own purpose: if it understood itself to be of singular importance, or of an importance that outweighed all others. Much of the failure of individual assaults on the web of domination derives from their incompleteness, and from the resulting blindspots that leaves domination able to renew, consolidate and express itself in a different, but related, form.[30]

Therefore, in our attempt to undermine the web we can follow the tradition of critique, as exemplified by social ecology. Human domination of nature can be systematically related to human domination of human, in terms of both form and content, without the former being reducible to the latter. Such an understanding would give us a partial, fragmentary understanding of the web. Similarly, we can pursue avenues of inquiry opened up by deep ecology. The fundamental strength of deep ecology lies in its recognition of the relational character of the self, the intrinsic connectedness of the self and its environment. Its fundamental weakness lies in the company it chooses to keep: American nature mysticism and Eastern religious/philosophical traditions such as Buddhism that have prevented it from developing this insight in a sufficiently rigorous or useful way. Again, deep ecology can broaden our reflective awareness of the structure of the web.

What of ecofeminism? When it is not engaged in some de-lusional attempt to establish male–female oppression as archetypal ur-domination, ecofeminism, as the systematic examination of distinct forms of domination and the relation between such forms, has developed certain ideas of significance. In particular, ecofeminists such as

Plumwood have argued, quite plausibly, that the various interwoven dualisms of Western culture – mind/body, male/female, reason/emotion, and subject/object – have interacted to form a similarly interwoven series of oppressions. These series of oppressions are linked by a common ideology, and this is the ideology of the control of reason over nature. In all forms of oppression, what the oppressed groups have in common is that they are all counted as part of the sphere of nature, and, as such, located outside the sphere of reason. It would be beyond present concerns to evaluate this claim. What is of present concern, however, is what to make of this claim should it be true. And, unfortunately, in the writings of all too many ecofeminists, the response has been, predictably but disappointingly, to reject reason. Reason is regarded as simply a tool of oppression, and therefore to be regarded as suspect on that basis. This flight from reason is, in many ways, of a piece with that exhibited by the deep ecology movement, where reason is regarded as a tool of environmental exploitation and to be rejected on that basis. A more appropriate response to the reason/nature dualism would be to break it down by showing that *reason is part of nature*. This is one the conclusions of Chapter 8, where it was argued that processes of rational inference have developed essentially in conjunction with processes of environmental manipulation, to the extent that such processes often contain environmental structures as constituents.

Demonstrating the worldly nature of all that has been thought to separate humans from worldly nature is, potentially, that which saves. The answer to a dichotomy and a value judgement predicated on that dichotomy is not to reverse the value judgement but to dismantle the dichotomy. That which is nature has been excluded from that which is reason, and systematically denigrated on that basis. The answer to this, contrary to some deep ecologists and some ecofeminists, is not to reject reason. The answer is to break down the dichotomy. But, crucially, it is *not* to break down the dichotomy in a way that fundamentally reinforces the value judgement predicated on that dichotomy. This, in effect, is the strategy of humanism, of neo-Kantian idealism, which breaks down the dichotomy between reason and nature by showing that nature is really a construction of reason. This reinforces the original value judgement by making nature ontologically and epistemologically dependent on reason, hence axiologically dependent on rational consciousness. To break down the reason/nature dichotomy and to do so in a way that does not reinforce the original value judgement predicated on that dichotomy, to show the worldly nature of reason: this is the fundamental task of environmental thought. And it is this task that requires a genuinely *environmental turn* in philosophy. Showing the need for such a turn, and indicating, in admittedly broad strokes, how such a turn might proceed, has been the task of this book.

Notes

1 'Where the Danger Is Grows also That which Saves'

1 Matt Ridley, *The Origins of Virtue* (London: Viking, 1996), p. 15.
2 J. E. Lovelock, *Gaia* (Oxford: Oxford University Press, 1979).
3 The expression 'deep ecology' was first coined by Arne Naess, 'The shallow and the deep, long range acology movements. A summary', *Inquiry*, 16 (1973), pp. 95–100.
4 Gilbert Ryle, *The Concept of Mind* (London: Hutchinson, 1949).
5 J. Baird Callicott, *In Defense of the Land Ethic* (New York: SUNY, 1989), p. 103.
6 See, for example, J. Baird Callicott, 'Intrinsic value, quantum theory, and environmental ethics' and 'The metaphysical implications of ecology', both reprinted in *In Defense of the Land Ethic*.
7 Martin Heidegger, 'The question concerning technology', in D. Krell (ed.), *Heidegger: Basic Writings* (New York: Harper & Row, 1977).
8 Friedrich Nietzsche, *Twilight of the Idols*, trans. R. J. Hollingdale (London: Penguin, 1968), pp. 40–1.
9 I am not claiming that Nietzsche understood the activity of willing in this way.
10 See, for example, Michael Dummett, *The Logical Basis of Metaphysics* (Cambridge, Mass.: Harvard University Press, 1991).
11 Eugene Hargrove, *Foundations of Environmental Ethics* (Texas: University of North Texas Press, 1989).
12 Holmes Rolston III, 'Values in Nature', *Environmental Ethics*, 3 (1981), pp. 113–28.
13 J. Baird Callicott, 'Intrinsic value, quantum theory, and environmental ethics', in *In Defense of the Land Ethic*, p. 167.
14 See Hilary Putnam, 'The meaning of "meaning"', in his *Mind, Language and Reality: Philosophical Papers*, Vol. 2 (Cambridge: Cambridge University Press, 1975). Also, Tyler Burge, 'Individualism and the mental', in *Midwest Studies in Philosophy*, 4 (1979), pp. 73–121.
15 See James J. Gibson, *The Senses Considered as Perceptual Systems* (Boston: Houghton-Mifflin, 1966); *The Ecological Approach to Visual Perception* (Boston: Houghton-Mifflin, 1979). The work in artificial intelligence I have in mind is known as connectionism or parallel distributed processing. See especially D. E. Rumelhart, J. M. McClelland and the PDP Research Group, *Parallel Distributed Processing*, Vol. 1 (Massachusetts: MIT Press, 1986). Putnam is discussed in Chapter 6; Gibson and PDP are discussed in Chapter 8.
16 Further attacks on neo-Kantian idealism/humanism can, in effect, be found in my *The Body in Mind: Understanding Cognitive Processes* (Cambridge: Cambridge University Press, 1999).

2 Intrinsic Value and Why (We Think) It's Needed

1 The scenario is due to Mary Midgeley, 'Duties concerning islands', in Robert Elliot (ed.), *Environmental Ethics* (Oxford: Oxford University Press, 1995). It is a version of the famous 'last man scenario' of Richard and Val Routley, 'Against the inevitability of human chauvinism', in K. Goodpaster and K. Sayre (eds.), *Ethics and the Problems of the 21st Century* (Indiana: University of Notre Dame Press, 1979), pp. 36–59.
2 Richard Sylvan, *Universal Purpose, Terrestrial Greenhouse, and Biological Evolution* (Canberra: Australian National University, 1990).
3 Janna L. Thompson, 'Preservation of wilderness and the good life', in R. Elliot and A. Gare (eds.), *Environmental Philosophy* (St Lucia: University of Queensland Press, 1983).
4 Classic statements of utilitarianism are to be found in Jeremy Bentham, *An Introduction to the Principles of Morals and Legislation*, ed. W. Harrison (Oxford: Oxford University Press, 1948); and J. S. Mill, 'Utilitarianism' in his *Utilitarianism and Other Writings*, ed. M. Warnock (New York: NAL, 1962). Influential recent treatments include, R. M. Hare, *Moral Thinking* (Oxford: Oxford University Press, 1961); S. Scheffler (ed.), *Consequentialism and its Critics* (Oxford: Oxford University Press, 1988); A. Sen and B. Williams (eds.), *Utilitarianism and Beyond* (Cambridge: Cambridge University Press. 1982); J. J. C. Smart and B. Williams, *Utilitarianism: For and Against* (Cambridge: Cambridge University Press, 1973).
5 For influential treatments of the issues raised by future generations, see B. Barry and R. I. Sikora (eds.), *Obligations to Future Generations* (Philadelphia: Temple University Press, 1978).
6 John Rawls, *A Theory of Justice* (Oxford: Oxford University Press, 1971). For an extension of neo-Rawlsian contractualism beyond the human sphere, see my *Animal Rights: a Philosophical Defence* (London: Macmillan, 1998); and also my 'Contractarianism and animal rights', *Journal of Applied Philosophy*, 14, 3 (1997), pp. 235–47.
7 For a discussion of some of the problems involved in incorporating future generations into moral theory, see Robert Elliot, 'The rights of future people', *Journal of Applied Philosophy*, 6 (1989), pp. 159–70. Also Barry and Sikora, *Obligations to Future Generations*.
8 Bryan Norton, *Toward Unity Amongst Environmentalists* (New York: Oxford University Press, 1991).
9 Warwick Fox, 'What does the recognition of intrinsic value entail?', *Trumpeter*, 10 (1993).
10 Peter Singer, *Animal Liberation* (London: Harper Collins, 1975).
11 Tom Regan, *The Case for Animal Rights* (London: Routledge, 1984).
12 Mark Rowlands, *Animal Rights* (London: Macmillan, 1998).
13 This is essentially the position I take in *Animal Rights*, where I focus, in particular, on the concept of equal consideration as it appears in human-based ethical theories.
14 For an influential discussion of the difference between a sentience based approach characteristic of animal rights literature and a genuine environment based approach, see Baird Callicott, 'Animal liberation: a triangular affair', in his *In Defense of the Land Ethic* (New York: SUNY, 1989).

15 Tom Regan, for example, writes: 'That an individual animal is among the last remaining members of a species confers no further right on that animal and its right not to be harmed must be weighted equitably with the rights of any others who have this right.' *The Case for Animal Rights*, p. 359.

16 This is a rehearsal of a point made by Tom Regan in a classic paper, 'The nature and possibility of an environmental ethic', *Environmental Ethics*, 3 (1981), pp. 19–34.

17 G. E. Moore, 'The conception of intrinsic value', in his *Philosophical Studies* (London: Routledge and Kegan Paul, 1922), p. 260.

18 Perhaps the principal objectivist defender of intrinsic value is Holmes Rolston III. See his *Philosophy Gone Wild: Essays in Environmental Ethics* (New York: Prometheus, 1989); *Environmental Ethics: Duties to and Values in the Natural World* (Philadelphia: Temple University Press, 1988); *Conserving Natural Value* (New York: Columbia University Press, 1994).

19 Two principal defenders of subjectivist accounts of intrinsic value are J. Baird Callicott, *In Defense of the Land Ethic* (New York: SUNY, 1989), and Robert Elliot, 'Intrinsic value, environmental obligation and naturalness', *The Monist*, 75, 2 (1992), pp. 138–60. Both will be discussed in Chapter 4.

3 Objectivist Theories of Intrinsic Value

1 Kenneth Goodpaster, 'On being morally considerable', *Journal of Philosophy*, 78 (1978), pp. 308–25. Paul W. Taylor, 'In defence of biocentrism', *Environmental Ethics*, 5 (1983), pp. 237–43, and *Respect for Nature: a Theory of Environmental Ethics* (Princeton: Princeton University Press, 1986). Holmes Rolston III, *Environmental Ethics: Duties to and Values in the Natural World* (Philadelphia: Temple University Press, 1988).

2 Peter Miller, 'Value as richness: toward a value theory for the expanded naturalism in environmental ethics', *Environmental Ethics*, 4 (1982), pp. 101–14.

3 G. E. Moore, *Principia Ethica* (Cambridge: Cambridge University Press, 1903).

4 Kenneth Goodpaster, 'On being morally considerable', *Journal of Philosophy*, 78 (1978), pp. 308–25.

5 Miller, 'Value as richness'.

6 J. Baird Callicott, 'Intrinsic value, quantum theory, and environmental ethics', in his *In Defense of the Land Ethic* (New York: SUNY, 1989), p. 159.

7 Holmes Rolston III, *Environmental Ethics: Duties to and Values in the Natural World*. Also, *Conserving Natural Value* (New York: Columbia University Press, 1994).

8 I develop this point at greater length in my *Animal Rights* (London: Macmillan, 1998). See also my 'Killing' (unpublished ms).

9 Rolston, *Environmental Ethics*, p. 101.

10 J. Baird Callicott, 'Intrinsic value in nature: a metaethical analysis', in his *Beyond the Land Ethic: More Essays in Environmental Philosophy* (New York: SUNY Press, 1999), pp. 239–61.

11 Rolston, *Environmental Ethics*.

12 Rolston, *Environmental Ethics*.

13 Rolston, *Conserving Natural Value*, p. 177.

4 Subjectivist Theories of Intrinsic Value

1 Robert Elliot, 'Intrinsic value, environmental obligation and naturalness', *The Monist*, 75, 2 (1992), pp. 138–60.

2 Elliot, 'Intrinsic value, environmental obligation and naturalness', p. 140.

3 Elliot, 'Intrinsic value, environmental obligation and naturalness', p. 140.

4 Elliot, 'Intrinsic value, environmental obligation and naturalness', p. 143.

5 Elliot, 'Intrinsic value, environmental obligation and naturalness', p. 146.

6 Elliot, 'Intrinsic value, environmental obligation and naturalness', p. 147.

7 Elliot, 'Intrinsic value, environmental obligation and naturalness', p. 147.

8 Callicott's more traditional approach to intrinsic value is to be found in 'The conceptual foundations of the land ethic', in *In Defense of the Land Ethic* (New York: SUNY, 1989), pp. 75–99; and 'Hume's is/ought dichotomy and the relation of ecology to Leopold's land ethic', *In Defense of the Land Ethic*, pp. 117–27. His more radical approach can be found in 'The metaphysical implications of ecology', *In Defense of the Land Ethic*, pp. 101–14; and 'Intrinsic value, quantum theory, and environmental ethics', *In Defense of the Land Ethic*, pp. 157–74.

9 The classic statement of Hume's meta-ethical theory is to be found in Book III of his *A Treatise of Human Nature*, ed. Pall S. Ardal (London: Fontana, 1962).

10 Darwin's account of the evolution of the social instincts is to be found in his *The Descent of Man* (London: Murray, 1871).

11 Also Leopold, *A Sand County Alamanac* (New York: Oxford University Press, 1949), pp. 203–4.

12 To stave off numerous misunderstandings of his work, Callicott is especially clear on the role and importance of a Leopoldian based cognitive redirection of natural sentiments in his latest book, *Beyond the Land Ethic* (New York: SUNY Press, 1999). See especially 'Just the facts, Ma'am' and 'Do deconstructive ecology and sociobiology undermine the land ethic?'

13 The charge is made by Warwick Fox, 'A postscript on deep ecology and intrinsic value', *Trumpeter*, 2, 4 (1985), pp. 20–3; and by K. S. Schrader-Frechette, 'Biological holism and the evolution of ethics', *Between the Species*, 6 (1990), pp. 185–92.

14 See his 'Can a theory of moral sentiments support a genuinely normative environmental ethic?' in *Beyond the Land Ethic* (New York: SUNY Press, 1999), pp. 99–116.

15 'Just the facts, Ma'am', in *Beyond the Land Ethic* (New York: SUNY Press, 1999), pp. 79–98.

16 'Just the facts, Ma'am', p. 88.

17 'How environmental theory may be put into practice', in *Beyond the Land Ethic*, pp. 45–58, 50.

18 Callicott, at least at one point, does acknowledge that his position is group selectionist. See *In Defense of the Land Ethic*, p. 118 n. He does not, however, seem to appreciate the seriousness of this admission.

19 Classic statements of the concept of an evolutionarily stable strategy are to be found in J. Maynard-Smith and G. R. Price 'The logic of animal conflicts', *Nature*, 246 (1973), pp. 15–18; and J. Maynard-Smith and G. A. Parker 'The logic of asymmetric contests', *Animal Behaviour*, 24, (1976), pp. 159–75.

20 The example is taken from Richard Dawkins, *The Selfish Gene* (Oxford: Oxford University Press, 1976), pp. 75–9.

21 See, for example, Matt Ridley, *The Origins of Virtue* (London: Viking, 1996).

5 Radical Approaches to the Value of Nature

1 Callicott's discussion of quantum theory is found in 'Intrinsic value, quantum theory, and environmental ethics'. His most complete discussion of ecological theory is in 'The metaphysical implications of ecology'. Both are found in *In Defense of the Land Ethic* (New York: SUNY 1989), pp. 157–74, 101–14 respectively.

2 'Intrinsic value, quantum theory, and environmental ethics', p. 166.

3 'Intrinsic value, quantum theory, and environmental ethics', p. 169.

4 'Intrinsic value, quantum theory, and environmental ethics', pp. 169–70.

5 'Intrinsic value, quantum theory, and environmental ethics', p. 170.

6 'Intrinsic value, quantum theory, and environmental ethics', p. 170.

7 The analogy is originally due to Harold Morowitz, 'Biology as a cosmological science', *Main Currents in Modern Thought*, 28 (1972), p. 156.

8 Arne Naess, 'The shallow and the deep, long range ecology movement: a summary', *Inquiry*, 16 (1973).

9 'The metaphysical implications of ecology', pp. 110–11.

10 'The metaphysical implications of ecology', pp. 101–2. See also, 'Intrinsic value, quantum theory, and environmental ethics', p. 166.

11 Paul Shephard, 'Ecology and man: a viewpoint', in P. Shephard and D. McKinley (eds.), *The Subversive Science: Essays Towards and Ecology of Man* (Boston: Houghton-Mifflin, 1967), p. 3. Quoted from Callicott, 'The metaphysical implications of ecology', p. 113.

12 'The metaphysical implications of ecology', p. 114.

13 See also, Kenneth Goodpaster, 'From egoism to environmentalism', in K. Goodpaster and K. Sayre (eds.), *Ethics and the Problems of the 21st Century*, (Notre Dame: Notre Dame University Press, 1979), pp. 21–35.

14 Ruth Millikan, 'What is behaviour, part II: the green grass growing all around', Millikan, *White Queen Psychology and Other Essays for Alice* (Massachusetts: MIT Press, 1993).

15 'What is behaviour, part II', p. 159.

16 'What is behaviour, part II', p. 161.

17 For Millikan's wider account of organisms, minds, language, and so on, see the other essays in *White Queen Psychology*, and also her brilliant *Language, Thought and Other Biological Categories* (Massachusetts: MIT Press, 1984).

18 A similar point is made by Val Plumwood, 'Nature, self, and gender; feminism, environmental philosophy, and the critique of rationalism', *Hypatia*, 6 (1991), pp. 3–27.

6 Against Humanism I: Externalism

1 Heidegger's interpretation of Nietzsche as the culmination or destiny of Western metaphysics is propounded in his *Nietzsche* (Harper & Row, 1979).

For an alternative interpretation, see Walter Kaufmann, *Nietzsche* (New Jersey: Princeton University Press, 1974).

2 The story is an outline of an argument first presented in Hilary Putnam's classic paper 'The meaning of "meaning"', in K. Gunderson (ed.), *Language, Mind, and Knowledge: Minnesota Studies in the Philosophy of Science*, 7 (Minneapolis: University of Minnesota Press, 1975). For a similar, and equally classic account, see Tyler Burge, 'Individualism and the mental', in P. A. French, T. E. Uehling and H. K. Wettstein (eds.), *Midwest Studies in Philosophy*, 4 (Minneapolis: University of Minnesota Press, 1979). For a good account of externalism and its philosophical ramifications, see Colin McGinn, *Mental Content* (Oxford: Basil Blackwell, 1989).

3 Actually, they would not, of course, live in the same house as you, but in a duplicate of yours. They would read duplicate books. In other words, 'same' in the above passage means 'exactly similar' rather than 'identical'. Less obviously, but more importantly, they would have the same experiences as you only if these are non-intentionally specified; i. e. only if experiences are, here, assimilated to purely phenomenological states.

4 This point is made by McGinn, *Mental Content*, p. 10.

5 For useful distinctions between different forms of externalism and their compatibility or incompatibility with different forms of idealism, see McGinn, *Mental Content*, pp. 10–13.

6 Peter Strawson, *Individuals* (London: Methuen, 1959), pp. 30 ff.

7 Colin McGinn, *Mental Content*, p. 5.

8 Cynthia Macdonald, 'Weak externalism and mind-body identity', *Mind*, 99 (1990), pp. 387–404.

7 Against Humanism II: Evolution

1 The example, and the reference to Llinas, are both taken from Daniel Dennett, *Consciousness Explained* (London: Little Brown 1991), p. 177.

2 Richard Dawkins, *The Extended Phenotype* (Oxford: Oxford University Press, 1982), p. 42.

3 Dawkins, *The Extended Phenotype*, p. 47.

4 Dawkins, *The Extended Phenotype*, p. 200.

5 The example is taken from Andy Clark, *Microcognition* (Massachusetts: MIT Press, 1989).

6 J. C. Holmes and W. M. Bethel, 'Modification of intermediate host behaviour by parasites', in E. U. Canning and C. A. Wright (eds.), *Behavioural Aspects of Parasite Transmission* (London: Academic Press, 1972), pp. 123–49. W. M. Bethel and J. C. Holmes, 'Altered evasive behaviour and responses to light in amphipods harboring acanthocephalan cystacanths', *Journal of Parasitology*, 59 (1973), pp. 945–56.

7 W. Wickler, 'Evolution-oriented ethology, kin selection, and altruistic parasites', *Zeitschrift für Tierpsychologie*, 42 (1976), pp. 206–14. M. Love, 'The alien strategy', *Natural History*, 89 (1980), pp. 30–2.

8 Dawkins, *The Extended Phenotype*, p. 67.

9 Dawkins, *The Extended Phenotype*, p. 60.

10 Dawkins, *The Extended Phenotype*, p. 59. See also, R. Dawkins and J. R. Krebs, 'Animal signals: information or manipulation?', in J. R. Krebs and N. B. Davies (eds.), *Behavioural Ecology* (Oxford: Blackwell, 1978), pp. 282–309.

11 A similar claim is captured by Andy Clark in what he calls the '007 Principle'. See *Microcognition* (Massachusetts: MIT Press, 1989), chapter 4.

12 This chapter is a condensed version of an argument I first presented in my *The Body in Mind: Understanding Cognitive Processes* (Cambridge: Cambridge University Press, 1999), chapter 4.

8 Against Humanism III: Cognition

1 S. J. Gould and E. S. Vrba, 'Exaptation: a missing term in the science of form', *Paleobiology*, 8, 1 (1982), pp. 4–15.

2 *The Body in Mind: Understanding Cognitive Processes* (Cambridge: Cambridge University Press, 1999).

3 Actually, this is not quite right. If, as is generally accepted, information is a form of nomological dependence, then information is pretty much ubiquitous, and so would be carried by the processes involved in respiration. The point is, however, that this information is not relevant to the cognitive task at hand.

4 David Marr, *Vision* (San Francisco: W. H. Freeman, 1982).

5 For Gibson's ecological account of visual perception, see especially his *The Senses Considered as Perceptual Systems* (Boston: Houghton-Mifflin, 1966), and *The Ecological Approach to Visual Perception* (Boston: Houghton-Mifflin, 1979).

6 H. A. Sedgwick, 'The visible horizon: a potential source of information for the perception of size and distance', Ph. D. dissertation, Cornell University (1973).

7 Cole, Hood and McDermott, 'Ecological niche picking', in Ulric Neisser (ed.), *Memory Observed* (San Francisco: W. H. Freeman, 1982).

8 See my 'Against methodological solipsism: the ecological approach', in *Philosophical Psychology*, 8, 1 (1995), pp. 5–24. Also, *The Body in Mind*, p. 120.

9 A. R. Luria and L. S. Vygotsky, *Ape, Primitive Man, and Child* (Massachusetts: MIT Press, 1992).

10 Merlin Donald, *Origins of the Modern Mind* (Massachusetts: Harvard University Press, 1991), pp. 308–25.

11 The *locus classicus* of recent connectionist approaches to cognitive modelling is to be found in D. E. Rumelhart, J. L. McClelland and the PDP Research Group, *Parallel Distributed Processing: Explorations in the Microstructure of Cognition*, vol. 1, *Foundations* (Massachusetts: MIT Press, 1986). Good introductions are to be found in Andy Clark, *Microcognition* (Massachusetts: MIT Press, 1989), and William Bechtel and Adele Abrahamsen, *Connectionism and the Mind* (Oxford: Blackwell, 1991).

12 D. E. Rumelhart, P. Smolensky, J. L. McClelland and G. E. Hinton, 'Schemata and sequential thought processes in PDP models', in Rumelhart, McClelland and the PDP Research Group, *Parallel Distributed Processing*, chapter 14.

13 Bechtel and Abrahamsen, *Connectionism and the Mind*.

14 This issue will be dealt with properly in Chapter 9.

9   Towards a Post-Humanist Theory of Value

1   I shall henceforth ignore non-naturalistic objectivism. Future reference to objectivism should, therefore, be understood as reference to the naturalistic version.

2   Callicott, 'Intrinsic value, quantum theory, and environmental ethics', p. 161.

3   'Intrinsic value, quantum theory, and environmental ethics', p. 161.

4   This sort of point is made quite forcefully by Peter Singer, *The Expanding Circle: Ethics and Sociobiology* (Oxford: Oxford University Press, 1981).

5   James J. Gibson, *The Ecological Approach to Visual Perception* (Boston: Houghton-Mifflin, 1979).

6   See G. Orians, 'Habitat selection: general theory and applications to human behaviour', in S. Lockard (ed.), *The Evolution of Human Social Behavior* (Chicago: Elsevier, 1980), pp. 49–66.

7   J. D. Balling and J. H. Falk, 'Development of visual preference for natural environments', *Environment and Behavior*, 14 (1990), pp. 5–28. R. Ulrich, 'Aesthetic and affective response to natural environment', in I. Altman and J. F. Wohlwill (eds.), *Behavior and the Natural Environment* (New York: Plenum, 1983), pp. 85–125.

8   G. Orians and J. H. Heerwagen, 'Evolved responses to landscapes', in J. Barkow, L. Cosmides and J. Tooby (eds.), *The Adapted Mind* (New York: Oxford University Press, 1992), pp. 555–79.

9   'Evolved responses to landscapes', pp. 563–5. Note that the terms 'richness', 'coherence', 'integrity' and 'diversity' are not ones employed by Orians and Heerwagen.

10   N. K. Humphrey, 'Natural aesthetics', in B. Mikellides (ed.), *Architecture for People* (London: Studio Vista 1980), pp. 59–73.

11   The example is borrowed from Andrew Brennan, 'Ecological theory and value in nature', *Philosophical Inquiry*, 8 (1986), pp. 66–96.

12   F. E. Clements, *Research Methods in Ecology* (Lincoln, Nebr: University Publishing Company 1905), p. 265.

13   See H. S. Horn, 'Markovian properties of forest succession', in M. L. Cody and J. M. Diamond (eds.), *Ecology and Evolution of Communities* (Cambridge, Mass: Harvard University Press 1975).

14   See E. P. Odum, 'The strategy of ecosystem development', *Science*, 164 (1969), pp. 262–70.

15   I say 'simple' subjectivist account in order to distinguish it from, for example, the more sophisticated account of Callicott. According to Callicott's account also, failure to appreciate the value of the environment is, in part, a cognitive failure. I am in wholehearted agreement with Callicott on this point. I think his arguments ultimately fail, but for the sorts of reasons canvassed in Chapter 4.

10   Perspectives on the Environmental Crisis: Social Ecology, Deep Ecology and Ecofeminism

1   Murray Bookchin, *The Ecology of Freedom* (Palo Alto, CA: Cheshire Books 1982); *Remaking Society* (Boston: South End Press 1989); *The Philosophy of Social Ecology* (Montreal: Black Rose Books, 1990).

2 *Remaking Society*, p. 44.
3 *Remaking Society*, pp. 60–1.
4 *Remaking Society*, p. 13.
5 As told, among very many others, by Richard Dawkins, *The Selfish Gene* (Oxford: Oxford University Press, 1976).
6 Daniel Dennett, *Darwin's Dangerous Idea* (London: Penguin 1995), p. 48.
7 His 'theory' of evolution is presented in most detailed form in *The Philosophy of Social Ecology*.
8 *Remaking Society*, p. 33
9 'Shambalic' is an industry term and is not to be confused with 'shambolic' (although, it might be argued that there are certain conceptual connections between the two). 'Shambalic' is defined, roughly, as 'pertaining to an item suitable for publication with the Shambala press of Boston'.
10 Bill Devall and George Sessions, *Deep Ecology: Living as if Nature Mattered* (Salt Lake City: Gibbs Smith, 1985), p. ix.
11 *Deep Ecology*, p. 65.
12 Robert Kirkman has pointed out that this type of reactive opposition to Cartesianism simply repeats the pattern established by romanticism in the nineteenth century. See his 'The problem of knowledge in environmental thought', in R. Gottlieb (ed.), *The Ecological Community* (London: Routledge 1997), pp. 193–207.
13 Val Plumwood, 'Nature, self, and gender: feminism, environmental philosophy and the critique of rationalism', *Hypatia*, 6 (1991), pp. 10–16, 23–6. I agree with Plumwood that this tripartite distinction is useful. However, my specific criticisms of each component differ somewhat from hers.
14 Warwick Fox, 'Deep ecology: a new philosophy of our time?', *Ecologist*, 14 (1984), p. 7. Despite the placement of a quotation from Fox at this point, he should definitely be included in the more sophisticated end of the deep ecology camp.
15 John Seed, 'Anthropocentrism', Appendix E in Devall and Sessions, *Deep Ecology*, 243.
16 Aldo Leopold, *A Sand County Almanac* (New York: Ballantine Books 1966), p. 197.
17 Warwick Fox, *Towards a Transpersonal Ecology: Developing New Foundations for Environmentalism* (Boston: Shambala, 1990).
18 Arne Naess, 'The shallow and the deep, long range ecology movement: a summary', *Inquiry*, 16 (1973), p. 96.
19 *Remaking Society*, p. 10.
20 For examples of all that is bad about ecofeminism, see, for example, I. Diamond and G. F. Orenstein (eds.), *Reweaving the World* (San Francisco: Sierra Club Books, 1990); and J. Plant (ed.), *Healing the Wounds* (Philadelphia: New Society Publishers, 1989).
21 This is not simply male whingeing on my part. My attention was drawn to this sort of ecofeminist response by one of the more thoughtful ecofeminists, Deborah Splicer, 'Wrongs of passage: three challenges to the maturing of ecofeminism', in K. J. Warren (ed.), *Ecological Feminism* (London: Routledge, 1994), pp. 29–41.
22 Ynestra King, 'Healing the wounds: feminism, ecology, and the nature/culture dualism', in *Reweaving the World*, pp. 106–7.

23　Ynestra King, 'Feminism and the revolt of nature', *Heresies*, 13 (1990), p. 12.

24　Sharon Doubiago, 'Mama coyote talks to the boys', in *Healing the Wounds*, p. 43.

25　Vandana Shiva, *Staying Alive: Women, Ecology, and Development* (London: Zed Books, 1988).

26　It is no surprise, then, to discover that King, for one, has been heavily influenced by Bookchin.

27　For example, Deborah Splicer, 'Wrongs of passage: three challenges to the maturing of ecofeminism', in K. J. Warren (ed.), *Ecological Feminism* (London: Routledge, 1994), pp. 29–41.

28　Val Plumwood, 'The ecopolitics debate and the politics of nature', in K. J. Warren (ed.), *Ecological Feminism*, p. 80.

29　Ludwig Wittgenstein, *Zettel*, trans. G. E. M Anscombe, G. E. M. Anscombe and G. H. von Wright (eds.) (Oxford: Basil Blackwell, 1981).

30　Here my discussion is indebted to Plumwood, 'The ecopolitics debate and the politics of nature', pp. 81–2.

Index